Communicating Clearly about Science and Medicine

To Dan, Ollie and India

Communicating Clearly about Science and Medicine

Making Data Presentations as Simple as Possible ... But No Simpler

JOHN CLARE

GOWER

Published by
Gower Publishing Limited
Wey Court East
Union Road
Farnham
Surrey, GU9 7PT
England

Gower Publishing Company
Suite 420
101 Cherry Street
Burlington,
VT 05401-4405
USA

www.gowerpublishing.com

British Library Cataloguing in Publication Data
Clare, John.
  Communicating clearly about science and medicine : making data presentations as simple as possible – but no simpler.
  1. Communication in science. 2. Communication in medicine. 3. Lectures and lecturing. 4. Science in mass media.
  I. Title
  808.5'1'0245–dc23

Library of Congress Cataloging-in-Publication Data
Clare, John.
  Communicating clearly about science and medicine : making data presentations as simple as possible – but no simpler / by John Clare.
       p. cm.
  Includes bibliographical references and index.
  ISBN 978-1-4094-4037-6 (paperback) – ISBN 978-1-4094-4038-3 (ebook)
  1. Communication in science. I. Title.

  Q223.C536 2011
  501'.4–dc23

                                                                2011036096

ISBN 9781409440376 (pbk)
ISBN 9781409440383 (ebk)

Printed and bound in Great Britain by the MPG Books Group, UK

Contents

List of Figures

About the Author

John Clare is passionate about communicating science and medicine. He will talk about it to anyone who will listen, and listen to anyone who talks about it. He has worked with thousands of presenters since he founded LionsDen Communications in 1992. They range from introverted research scientists who emerged from the lab blinking into the spotlight to inspirational presenters who are world leaders in their fields.

He attends many of the major medical congresses around the world and has been responsible for helping speakers to prepare for every kind of talk from plenary presentations of ground-breaking science to satellite symposia attended by a small number of specialists.

He has worked with most of the world's large pharmaceutical, biotech and vaccines companies, helping them to clarify and present their messages to triallists, regulators, investors, payers and journalists. He also coaches executives who need to present their business cases to senior management.

He is a regular moderator and presenter at medical meetings around the world, and has hosted media briefings and news conferences about science and drug development in Europe, the USA, Asia and South America. He has prepared many hundreds of scientists and physicians for interviews with every type of media outlet from international news agencies and world-renowned newspapers and TV programmes to scientific journals.

He was a journalist in newspapers and TV for many years. During that time he was a reporter, producer and news editor. He worked in a number of leading media outlets based in London, including ITN and *The Daily Mail*. He was the executive producer for LionsDen of *Organ Farm*, a TV documentary series about xenotransplantation. The series revealed and explored efforts to breed transgenic pigs whose organs could be transplanted into humans in need of a replacement major organ. It won many prestigious awards including two 'Freddies' at the TimeInc Health Awards in New York.

He and LionsDen are retained as issues and crisis consultants by many large pharmaceutical firms and scientific organisations.

He has a Master's Degree in Mass Communication. His MA thesis, *Town Criers in the Global Village: An Investigation into the Newsmaking Processes of International TV News Agencies* is regularly referenced and quoted.

In the UK he was awarded the *Communiqué Judges Award for Outstanding Healthcare Communications* in recognition of his long commitment to communicating about healthcare. He is the chief executive of LionsDen Communications, the international communication coaching and training specialists based in London. He writes regularly for magazines and websites and is a regular blogger and Tweeter. His previous book *John Clare's Guide to Media Handling* was published by Gower in 2001 and has been translated into Chinese, Arabic and Urdu. He is also the co-author (with Jenny Bryan) of *Organ Farm*, the book of the TV series, and *Patents, Patients and Profits*, published by SCRIP special reports.

You can find out more about him here:

http://www.lionsdencommunications.com

Preface

This book is a practical guide for anyone who aspires to present medical or scientific data to their peers and colleagues, and to talk confidently about it in the media and with other non-specialists. It addresses the specific challenges of talking about complex science in an engaging way.

> *It teaches readers how to combine the accuracy of peer-reviewed science with the narrative skills of journalism.*

Spoken communication skills are now essential for success in science and medicine. It is no longer sufficient to be a good physician or scientist. Anyone who seeks to play an important role in their field must also be an excellent communicator. They must be able to present, explain and be interviewed about their research, confidently and memorably. They must also be prepared to talk about their work in informal situations and meetings. These may be with colleagues, business partners, potential investors, regulators, journalists and medical students.

This is not just the case for young professionals starting out on their career. Once experienced professionals have achieved key opinion leader status, they need advanced spoken communication skills to maintain it.

Written communication in the form of a peer-reviewed publication has been the cornerstone of research and its dissemination since the seventeenth century, so why are verbal communication skills so important today? Because the way science is communicated has changed. The days of a publication appearing in a learned journal to be discussed solely by fellow experts and commented on weeks or months later have disappeared. The timescale is now compressed.

Key studies are now presented at major congresses and simultaneously published online. The lead authors are interviewed for congress TV channels and websites within minutes of their presentation ending. They then enter the lion's den of the news conference, to share their findings with journalists from all over the world. Once the news spreads, they will be besieged with requests for interviews and talks about their work. They will be called upon to explain, contextualise and defend their findings by scientists, non-scientists, funders, regulators, journalists of different types, pressure groups and many other interested parties.

These are the overlapping rings of science communication: Publication, presentation and interviews. This book equips scientists, physicians and others involved in the scientific community to deal with the last two.

By nature the medical and scientific worlds are complex and speaking about them is challenging. It is not the same as talking about the latest mobile phone, breakfast cereal or fashion range. The challenge for all communicators is to say things in a way which cannot be misunderstood. Doing so in the field of science and medicine is more difficult than most areas, but has never been more important. This is the case whether you are making a plenary presentation at a major congress or taking advantage of a chance meeting in the staff restaurant or by the water cooler. In today's ever more complex, fast-changing world, every interaction counts. Whether your topic involves complicated epidemiology, a novel mechanism of action or a breakthrough in understanding the pathophysiology of a disease, it needs to be communicated clearly. The key is to convey the right message in the right language to suit the audience.

The advice attributed to Einstein is a great starting point:

Make things as simple as possible, but no simpler.

In the medical and scientific field the challenges of clear communication are great, the risks of getting it wrong are high, and the consequences of doing so can be hugely expensive, in terms of credibility and wasted opportunity and investment. A mishandled presentation to regulators such as the Food and Drugs Administration (FDA) or the European Medicines Agency (EMA) can set a project back months or years. Failing to convey accurately the potential risks, benefits and potential of an investigational compound can lead to legal action from investors. The stakes are high.

Achieving recognition as a communicator is increasingly important for the career progression of physicians and scientists. Universities and institutions are ranked according to the amount and calibre of their research, and successful researchers attract more funding for further work. Presenting and publicising their research, in addition to having it published in respected scientific journals, is a crucial element of this process.

Undertaking speaking and publicity engagements requires skills which are not taught routinely as part of normal scientific training. For some researchers, communication has traditionally been something left to others. Until recently, many of them shied away from getting involved. That position is no longer sustainable. Every university or scientific organisation needs its leading researchers to communicate clearly to a range of audiences.

Over the last 20 years I have worked with some of the world's leading medical and scientific thought leaders, presenters and interviewees. With this book I have tried to put the benefit of that experience in your hands and take you on a journey from complexity to clarity. I hope you find it engaging, enjoyable and useful.

I would be delighted to receive your comments and suggestions to:

john.clare@lionsdencommunications.com

John Clare

Acknowledgements

Helping doctors, scientists and pharmaceutical executives improve their spoken communication skills is an enormously enjoyable and fulfilling way of earning a living. It's also a privilege. I started my business in 1992 after a successful career in print and broadcast journalism. Since then I have had the pleasure of working with some of the world's leading scientists, academics, researchers and business visionaries. Just as I have tried to help them to communicate, they have tried to teach me a little about science and medicine.

From both points of view the experience has been exhilarating, rewarding and at times frustrating. The journey from complexity to clarity can be hard work as we try to maintain the integrity of the science while removing any barriers to clear understanding.

Over the years I have been helped by more people than I could possibly acknowledge here. Some are leaders in their fields who have produced brilliant examples of how to communicate complex science. Others have suggested particular techniques for improving communication skills, or have commented on what they have found helpful in my own teaching. This, coupled with my own enthusiasm and exploration of the very best ways of communicating science, has produced the body of techniques, experience and anecdotes you will find in this book.

In particular I would like to acknowledge the help from all the LionsDen team who work with me in the UK and around the world. It is hard to imagine a more talented team of individuals who work so successfully on a daily basis to communicate such complex raw material in so many languages.

I would like to thank Lloyd Bracey for so many stimulating discussions, and for allowing me to use the grid system of presentation planning which he developed. You can see it in Chapter 4.

List of Abbreviations

ACC	American College of Cardiology
ACE	Angiotensin Converting Enzyme
AHA	American Heart Association
AIs	Aromatase Inhibitors
AIDS	Acquired Immune Deficiency Syndrome
ARB	Angiotensin Receptor Blocker
ARV	Antiretroviral (Drug)
ASCOT	Anglo Scandinavian Outcomes Trial
ASH	American Society of Haematology
BMJ	*British Medical Journal*
BP	Blood Pressure
BR	Background Regimen
CI	Confidence Interval
CFUs	Colony-Forming Units
CNN	Cable News Network
CTA	Call to Action
CV	Cardiovascular
DNA	Deoxyribonucleic Acid
DVT	Deep Vein Thrombosis
EBM	Evidence Based Medicine
ECG	Electrocardiograph/Electrocardiogram
EMA	European Medicines Agency
EQ	Emotional Quotient
ERs	Emergency Rooms
ERS	European Respiratory Society
ESC	European Society of Cardiology
FDA	Food and Drugs Administration
GM	Genetically Modified
HDL	High Density Lipoprotein
HIV	Human Immuno-Deficiency Virus
hsCRP	High Sensitivity C Reactive Protein

IMRAD	Introduction, Methods, Results and Discussion
INR	International Normalised Ratio
IQ	Intelligence Quotient
IV	Intravenous
LDL	Low Density Lipoprotein
LFT	Liver Function Test
MALES	Message, Audience, Language, Examples, Summary
MI	Myocardial Infarction
MIC	Minimum Inhibitory Concentration
MMR	Mumps, Measles, Rubella
MRSA	Methicillin-resistant Staphylococcus aureus
MS	Multiple Sclerosis
PEP	Point, Evidence, Point
PI	Principal Investigator
PLATO	Platelet Inhibition and Patient Outcomes
PPI	Proton Pump Inhibitor
RAM	Resistance Associated Mutations
RRMS	Relapsing Remitting Multiple Sclerosis
SSRIs	Selective Serotonin Reuptake Inhibitors
TED	Technology, Entertainment, Design
THR	Total Hip Replacement
TKR	Total Knee Replacement
TLAs	Three Letter Acronyms
T-scores	Bone Mineral Density (BMD) test
VTEs	Venous Thromboembolisms

Introduction: About this Book

Question: There are many books on communicating clearly, so why is this one different?

Answer: It's aimed at doctors and scientists, and anyone else who has to present or talk about clinical trials, science or medicine to a range of audiences. Scientific researchers, physicians who conduct clinical trials, academics, executives in pharmaceutical companies and representatives of scientific will find it invaluable.

Scientific communication poses its own challenges. The subject matter is complex and often requires the audience to have a certain level of prior knowledge to understand it correctly. Describing hazard ratios, interpreting Kaplan Meier curves and explaining confounding factors is different from talking about a new car or clothing range. Processes, for example in clinical trials, are laborious and tedious. Knowing how much of the detail to include and exclude requires judgement. Conclusions are rarely clear cut, and are often a matter of interpretation rather than hard facts. Communicating statistical risk and probability is challenging, especially to non-statisticians and non-scientists such as journalists. This book will look at these and many more challenges, then introduce powerful techniques for overcoming them.

It focuses on three types of activities:

1. **Peer to peer communication**, where you are talking to (and answering questions from) informed or specialist audiences.
2. **Onward communication**, where you need to communicate complex matters to non-experts, for example funding bodies or the public.
3. **Media interviews**, where you are required to tell a complex scientific story in a range of media, including newspapers, magazines, specialist journals, TV, radio and online.

Imagine you are one of the leading researchers on a trial of a promising new treatment for one of the major killers which attracts worldwide attention, such as diabetes, cancer, dementia or HIV/AIDS. You and your team have received the data, and the preliminary results are striking. You have been invited to present them at a satellite meeting at the year's major conference in the US.

You expect there will be a few hundred physicians, scientists and researchers in the audience, as well as medical journalists from around the world. *The New England Journal of Medicine* will simultaneously publish the findings online. The presentation will be a high profile event which has handed you a professional responsibility (to get it right) and a personal opportunity (to raise your profile in the medical community).

The findings are unclear, despite the smallest hint of a potential minor safety concern. If they are confirmed by larger and longer trials, this treatment could be what physicians and patients around the world have been waiting for. You want to do justice to the data, and strike the right balance in communicating the potential risks and benefits of the new treatment. You need to convey the promise it offers, while avoiding the 'miracle cure' and 'medical breakthrough' headlines you know the journalists will demand.

After your presentation, you will have to answer questions from scientific colleagues (and rivals!). Immediately following that there will be a series of interviews with scientific and medical correspondents accredited to the conference. They will file their stories and you will then be contacted by non-specialist journalists from all over the world, whose understanding of the subject matter may be sketchy at best. When you return to your institute you will be besieged with requests to talk about your data.

All of these situations require sophisticated communication skills. In particular, you will be dealing with different levels of understanding and prior knowledge.

Before any of that, you have to write the presentation. Where do you start? Where do you end? The data set is large and complex ... how much should you include? How much emphasis should you give to that potential small safety concern? How will you illustrate the key points, ensuring that every slide enhances the audience's understanding? When you take the stage, how can you ensure you will deliver an assured, impactful and memorable performance? How confident will you feel handling questions from internationally-renowned researchers with big reputations and even bigger egos? If that goes well, how can you handle the media interviews, balancing the hope and the hype? Then there will be countless other conversations about it, and your opinion will be sought regarding the larger trials.

This book is your starting point for success. It will teach you how to combine the accuracy of peer-reviewed science with the narrative skills of journalism.

It will help you develop and deliver impactful presentations on medical and scientific data. It will help you to move beyond PowerPoint and tell a clear, compelling story based on your data. It will show you how to develop clear messages and themes from complex data, while adhering to the advice attributed to Einstein: 'Make things as simple as possible ... but no simpler.'

It will also help you to plan for the media interviews, avoiding the traps which are waiting for unsuspecting scientists in the harsh glare of the media spotlight. It will use real-life examples of scientific media coverage to help you understand the media world, and to tailor your messages and activities to it. In addition, it will help you to handle all those other conversations you need to have, with colleagues, other triallists, pharmaceutical company executives, regulators and patient groups.

My Own Experience

I have spent nearly 20 years working with some of the world's leading medical and scientific thought leaders, helping them to prepare for crucial presentations at congresses, conferences and meetings. I have trained and coached thousands of physicians, scientists and pharmaceutical executives in presentation, communication and media skills in Europe, the US, South America, Russia and the Far East. This book is based on the accumulation of that expertise.

I understand the challenges of talking about science and medicine, and the responsibility of the task. I recognise the conflict between saying something new and exciting which will gather recognition, not to mention further research grants, and the need to avoid charges of bias by over-simplification or over-extrapolation of the findings.

I particularly enjoy coaching presenters and interviewees who have to present in English when it is not their first language, and the book includes much advice for you.

I use many real-life examples and illustrations throughout the book. Some are from very recent studies, while others are from landmark trials that changed medical practice. They all contain great lessons for anyone hoping to communicate science and medicine clearly.

In today's world, it is no longer sufficient to be a good doctor or scientist. You also need to be a skilled communicator. I hope this book gives you that opportunity.

1

Science Communication in the Twenty-First Century

The importance of the audience – suspicion about science – why talking to non-scientists is important – the overlapping rings of science communication – the convergence of communication channels – examples of great science communication.

Appreciate the Audience

If I had a magic wand and was able to wave it over every person in the world preparing a presentation, I would make them all do this: Think about the audience. Put yourself in their shoes and stand where they are standing. Only then can you see your situation from their point of view. From there you can develop a presentation or an informal talk that takes into account their level of knowledge, interest and expectations.

Let's do this now, and think of the audience in terms of the big picture. When we talk about science communication, we refer to a situation where an expert (you) communicates to non-experts (your audience). Their level of 'non-expertness' will vary hugely. Examples from opposite ends of the spectrum might be a presentation to fellow physicians and researchers at a medical congress, where you and they both understand the background, and where what you are telling them is the result of your latest specialist research, versus a television interview about your research on a show aimed at the general public.

In both instances, and every situation in between, our aim is clear spoken communication. The challenges, however, are very different. They start with the audience. Scientific colleagues start with a good level of understanding about your subject, and are ready to pick on the smallest details of inconsistency, error or lack of rigour. On the other hand, the public start with a suspicion about science, a fear of its power and often a feeling that it is too difficult to understand.

Ben Goldacre, author of *Bad Science*, his best-selling book based on his weekly column in *The Guardian* newspaper, has a theory about why the public is suspicious of science. He says that until the 1960s, if you learned about science at school, you could broadly understand how things worked … engines, powered flight, TV sets, space rockets and other inventions which were regarded as 'cutting edge' at the time. Since then, however, science has become so complicated that only real specialists can understand it. If you stopped someone in the street, would they be able to explain how a mobile phone works? Or a computer? Or satellite navigation? 'Something to do with microwaves, and radiation … and satellites,' is as close as most people get. Can you explain how the human genome was cracked? How Tyrosine Kinase Inhibitors work? As a scientist you may know the answers, but the public doesn't. What we don't understand makes us fearful. On top of that there have been so many 'scares about science' in recent years including fears over the safety of genetically modified (GM) crops, mobile phones linked to brain cancer, leukemia clusters around power lines and many other issues. If you have the public as your audience, you need to take this into account when you start to plan your presentation. Hence my advice: put yourself in their shoes. We will explore this in detail later in the book.

Given the difficulties, you may ask why we should communicate about science at all to non-scientists. Why shouldn't we confine our discussion to scientific journals, and only present to specialist colleagues? For centuries, that's exactly what happened, in most cases. Scientific research was conducted in a metaphorical black box. Scientists beavered away in secret, and when they had invented something which they believed was worthwhile, or maybe just clever, they handed it over to a grateful public. The public, when it thought about science at all, assumed it was conducted by bald men in white coats ('eggheads') who would run down a corridor shouting 'Eureka!' when they discovered something. They accepted the fruits of the research unquestioningly, and the scientists went back to their work.

Eventually the recipients started to ask questions: Is this the kind of development we want? Who decided what the scientists should invent? Is the new technology safe, and for how long? What if it falls into the wrong hands? Is it a good financial investment? Over time, these questions became criticisms, some justified, some not. The scientists began to discover the truth of C. Northcote Parkinson's quote about the consequences of failing to communicate: 'The void created by the failure to communicate is soon filled with poison, drivel and misrepresentation.'

The next step was that the public – and governments – demanded a stake in the planning of research, and a say in deciding what was ethically acceptable and financially justifiable. 'We're paying for this research out of public funds, so we want to know what's going on.' Nowadays, not communicating with non-scientists is not an option. Doctors and scientists are regularly called upon to explain and justify their work. Scientific topics such as climate change, energy use, cloning, stem cell research, the perceived risk of dangerous emissions from electricity pylons and mobile phones, risk of deep vein thrombosis (DVTs) on long haul flights and cancer risks from almost anything as well as the risks and benefits of vaccines and medications are high on the list of people's concerns. The way to allay those concerns is by communicating clearly about science.

The Overlapping Rings

Science communication takes place in a number of ways. The three most important are:

1. publication in peer-reviewed journals
2. presentations
3. media interviews

Within these, we can identify sub-sections:

1. Publication in peer-reviewed journals
   – often involves editorial comment in the journal
   – includes online publication on journal website

2. Presentations
   – to colleagues
   – to public

3. Media interviews
   – to specialist media
   – to general media
   – one-to-one interviews or press conferences

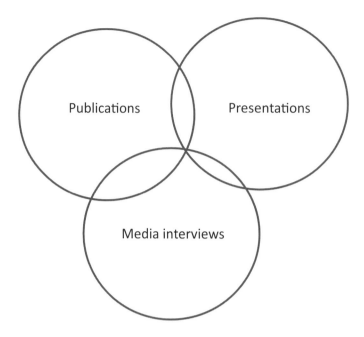

Figure 1.1   The overlapping rings of science communication

This book is concerned with the last two, which demand skills in spoken communication. However, presenting this as a list gives an inaccurate impression, as it suggests three discrete activities taking place sequentially on a continuum. In reality, they overlap. In the process, the audience changes too, from specialists to the general public. Today, I think of most science communication as three overlapping rings:

Presenting all three rings as the same size suggests that they are all equally important. In my view, this is increasingly the case, although without the peer-reviewed publication ring there would be nothing happening in the other two. Once you have achieved the publication, however, the other two take on equal significance with it among different but equally important audiences.

THE RINGS IN PRACTICE

As an example, consider a large study presented at the annual congress of the European Society of Cardiology (ESC). This is the largest medical meeting in Europe, and the largest cardiology meeting in the world. It is attended by approximately 25,000 physicians, 750 journalists and 5,000 others, including specialist nurses, scientific researchers, pharmaceutical executives, investors, equipment suppliers and exhibition staff. In 2009, it was held in Barcelona, Spain. On 29 August the results of a major study were presented. The PLATO trial was an investigation of a new anti-platelet agent aimed at preventing dangerous blood clots and reducing stroke, heart attacks and cardiovascular deaths in people with Acute Coronary Syndrome. It compared a new experimental drug, ticagrelor, with clopidogrel, a well-established medication.

The study was presented in the plenary session by Professor Lars Wallentin, Director of the Uppsala Clinical Research Centre, Sweden, and co-chair of the PLATO executive committee. It was standing room only in the packed plenary hall as more than a thousand physicians, researchers, pharmaceutical executives, investors and journalists crowded in to see the long-awaited results unveiled. Many attendees, including me, were forced to sit on the floor in the aisles. The presentation was broadcast live on the internet and immediately made available around the world. There were many more in the overspill hall and others watching on Congress TV around the conference centre.

The results were simultaneously published online in the *New England Journal of Medicine.* This allowed fellow specialists to delve into the details and comment on the study. Traditionally, when the scientific journals were print-only publications, comments took weeks or months to emerge. Now it happens online within minutes of the paper appearing on the journal's website.

As is usually the case, Professor Wallentin was allowed only five minutes to present the findings. This in itself posed a major challenge with a study which included more than 18,500 patients treated for up to 12 months. Deciding what to include and omit was a crucial decision. As I said in the introduction to this book, he needed to combine 'the accuracy of peer-reviewed science with the narrative skills of journalism'. He did an exemplary job. His presentation was followed by a five-minute commentary from 'The Discussant', another expert who gave his view of the study's strengths, weaknesses and relevance.

Journalists following the presentation in the hall were also following the online comments from specialists, and noting the comments from The Discussant. They were writing their stories based on a combination of the sources. The first headlines appeared on major medical news sites before Professor Wallentin had finished speaking.

After the plenary presentation, the professor and colleagues were interviewed about the study on Congress TV. You can find the interviews on the ESC's *YouTube* channel.

The next step was the press conference, attended by many of the 745 journalists accredited to the congress. They included news agency reporters, and correspondents from a wide range of publications from specialist to general, for example, *Cardiology Today* and *Heart.org* to *The New York Times*, *El Pais* or *The South China Morning Post*. The language at the press conference was less technical than that used in the plenary presentation. The key points needed to be clear, and relevance to the journalists' audience had to be explained.

This prompted follow-up calls from journalists all over the world. It is impossible to check how many stories appeared as a result, but a Google search for 'PLATO ticagrelor publication' produces 44,000 results.

This is a perfect illustration of the three overlapping rings, and how new science is communicated to scientists and non-scientists early in the twenty-first century.

Great Communicators

I have hosted many of these press conferences, and have experienced some outstanding examples of clear communication. One came from Professor Salim Yusuf, an internationally renowned professor of cardiology from McMaster University in Ontario, Canada. He was expected to speak for 15 minutes on the programme to make a presentation about a major new study at the American College of Cardiology (ACC) in Chicago. The ACC rules prevent anybody from holding an event at a time which may conflict with the ACC meeting. As a result, the press conference was held at 6.30 a.m. The results were so significant that more than a hundred journalists turned up.

Never a man to waste time or words, the professor said, 'I'm not going to bore you with a presentation. This was a big study, and it involved two drugs which can both reduce your blood pressure. The main question was: 'Is an Angiotensin Receptor Blocker (ARB), a newer drug, just as effective as a proven ACE (Angiotensin-Converting Enzyme) inhibitor, an older drug that is known to save lives, at preventing heart attacks and strokes? The answer is "Yes". We've proven it with a very high degree of confidence. The good news is that the ARB is slightly better tolerated so if you can't tolerate an ACE inhibitor, then we now know we can use the ARB with confidence. This gives doctors and patients an important choice. Now … questions?'

When faced with such a clear example of communication, there was little chance that the journalists could misunderstand the result or its significance, and the story made headlines around the world.

The two stories above demonstrate another fact about the overlapping rings: They are converging. The area of overlap in the diagram I used above was quite small. Thanks to the dominance of the internet and the ubiquitous nature of instant communication, the boundaries between the publication, presentation and media elements are becoming increasingly indistinct. The following diagram is probably a more accurate representation:

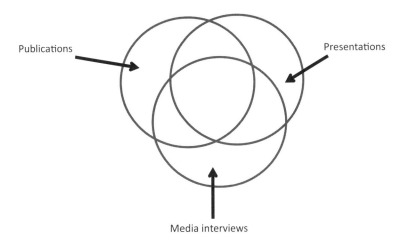

Publications

Presentations

Media interviews

Figure 1.2   Converging overlapping rings

Mastering the overlapping rings has a number of benefits: Public, professional and personal.

PUBLIC

The public gets to know about the latest developments in science and medicine. This may be relevant to their own lives, diseases and problems. Being informed about the latest scientific developments is a crucial part of public debate.

PROFESSIONAL

Exposure of this type raises the profile of the institution to which the presenter is affiliated. Universities and institutions are ranked according to the amount and calibre of their research, and successful researchers attract more funding for further work. Presenting and publicising their research, in addition to having it published in respected scientific journals, is a crucial element of this process.

PERSONAL

Achieving recognition as a communicator is increasingly important for the career progression of physicians and scientists. Spoken communication skills are the new cornerstone for success in science and medicine. It is no longer sufficient to be a good physician or scientist. Anyone who aspires to be a key player in their field must also be an excellent communicator. In particular, they must be able to present, explain and be interviewed about their research, confidently and memorably.

This is not just the case for young professionals starting out on their career. Once seasoned professionals have achieved Key Opinion Leader status, they need advanced spoken communication skills to maintain it.

Nobel Prize Standard Communication

I started this chapter by urging you to consider the audience. To see world class examples of an organisation which does exactly that, look up the announcements of the Nobel Prizes for science. Given the complexity of the science involved, the subject matter is necessarily complicated. However, the Nobel Prize organisation does a fabulous job of making it understandable.

The press releases announcing the prizes are a model of clarity. The Nobel organisation issues different versions of background reading to explain the research which has been honoured. It supplies background material aimed at scientists, which it calls *advanced information* and non-scientists, known as *public information*.

In the words of the opening chapter of this book, the Nobel Prize organisation really does aim to follow the advice attributed to Einstein to 'Make things as simple as possible, but no simpler.' They also achieve what I am aiming to do with this book, and combine the accuracy of peer-reviewed science with the narrative skills of journalism.

Here is an extract from the general press release announcing the 2010 Nobel Prize for Physics:

> *A thin flake of ordinary carbon, just one atom thick, lies behind this year's Nobel Prize in Physics. Andre Geim and Konstantin Novoselov have shown that carbon in such a flat form has exceptional properties that originate from the remarkable world of quantum physics.*
>
> *Graphene is a form of carbon. As a material it is completely new – not only the thinnest ever but also the strongest. As a conductor of electricity it performs as well as copper. As a conductor of heat it outperforms all other known materials. It is almost completely transparent, yet so dense that not even helium, the smallest gas atom, can pass through it. Carbon, the basis of all known life on earth, has surprised us once again.*
>
> *Geim and Novoselov extracted the graphene from a piece of graphite such as is found in ordinary pencils. Using regular adhesive tape they managed to obtain a flake of carbon with a thickness of just one atom. This at a time when many believed it was impossible for such thin crystalline materials to be stable.*
>
> *However, with graphene, physicists can now study a new class of two-dimensional materials with unique properties. Graphene makes experiments possible that give new twists to the phenomena in quantum physics. Also a vast variety of practical applications now appear possible including the creation of new materials and the manufacture of innovative electronics. Graphene transistors are predicted to be substantially faster than today's silicon transistors and result in more efficient computers.*
>
> *Since it is practically transparent and a good conductor, graphene is suitable for producing transparent touch screens, light panels, and maybe even solar cells.*[1]

Here are excerpts from two documents explaining the background to the 2010 Nobel Prize for Chemistry, awarded to three scientists for the development of palladium-catalyzed cross coupling. Notice the difference in language and complexity of ideas:

---

1 'The 2010 Nobel Prize in Physics – Press Release.' Nobelprize.org. 18 Jun 2011 http://nobelprize.org/nobel_prizes/physics/laureates/2010/press.html

General Press Release

GREAT ART IN A TEST TUBE

Organic chemistry has developed into an art form where scientists produce marvellous chemical creations in their test tubes. Mankind benefits from this in the form of medicines, ever-more precise electronics and advanced technological materials. The Nobel Prize in Chemistry 2010 awards one of the most sophisticated tools available to chemists today.

This year's Nobel Prize in Chemistry is awarded to Richard F. Heck, Ei-ichi Negishi and Akira Suzuki for the development of palladium-catalyzed cross coupling. This chemical tool has vastly improved the possibilities for chemists to create sophisticated chemicals, for example carbon-based molecules as complex as those created by nature itself.

Carbon-based (organic) chemistry is the basis of life and is responsible for numerous fascinating natural phenomena: colour in flowers, snake poison and bacteria-killing substances such as penicillin. Organic chemistry has allowed man to build on nature's chemistry; making use of carbon's ability to provide a stable skeleton for functional molecules. This has given mankind new medicines and revolutionary materials such as plastics.

In order to create these complex chemicals, chemists need to be able to join carbon atoms together....[2] Here is the introduction to the 'public information' section:

A Powerful Tool for Chemists

There is an increasing need for complex chemicals. Humanity wants new medicines that can cure cancer or halt the devastating effects of deadly viruses in the human body. The electronics industry is searching for substances that can emit light, and the agricultural industry wants substances that can protect crops. The Nobel Prize in Chemistry 2010 rewards a tool that has improved the ability of chemists to satisfy all of these wishes very efficiently: palladium-catalyzed cross coupling.

---

2   'The Nobel Prize in Chemistry 2010 – Press Release.' Nobelprize.org. 18 Jun 2011 http://nobelprize.org/nobel_prizes/chemistry/laureates/2010/press.html

By contrast, here is an excerpt from the 'scientific background':

> There are two types of cross-coupling reactions according to this principle that have become important in organic synthesis. These two types of reactions are shown in equations 1 and 2.
>
> Both reactions are catalyzed by zerovalent palladium and both reactions employ an organohalide RX (or analogous compound) as the electrophilic coupling partner. However, the nucleophilic coupling partner differs in the two reactions. In the first type (eq. 1) it is an olefin whereas in the second type (eq. 2) it is an organometallic compound R''M.

I hope you can see from these examples how targeting different audiences is so important for clear scientific communication.

OTHER EXAMPLES OF GREAT COMMUNICATION

Whatever audience you have, you can find great examples of excellent communication on the internet. I recommend you take a look at what's available on your own specialist subject, and see if you can do better. Patient websites or sites like http://Webmd.com are an excellent place to start. As an example of communication at both ends of the expert–non-expert spectrum, here are two descriptions of how abiraterone, a treatment in development for castration-resistant prostate cancer, works:

Scientific Version from Wikipedia:

> Abiraterone inhibits 17 α-hydroxylase/C17,20 lyase (CYP17A1), an enzyme which is expressed in testicular, adrenal, and prostatic tumor tissues. CYP17 catalyzes two sequential reactions: (a) the conversion of pregnenolone and progesterone to their 17-α-hydroxy derivatives by its 17 α-hydroxylase activity, and (b) the subsequent formation of dehydroepiandrosterone (DHEA) and androstenedione, respectively, by its C17,20 lyase activity. DHEA and androstenedione are androgens and precursors of testosterone. Inhibition of CYP17 activity by abiraterone thus decreases circulating levels of testosterone.

Patient Version from Cancer Research UK:

Abiraterone is a new hormone therapy drug that researchers are looking into for prostate cancer. It is also called CB7630 or abiraterone acetate. It works in a different way to other hormone treatments for prostate cancer.

The male hormone testosterone stimulates prostate cancers to grow. Stopping the body making testosterone can slow the growth of the cancer, or even shrink it. Most testosterone is made by the testes but a small amount is made by other tissue in the body including the cancer itself. To make testosterone the body needs an enzyme called cytochrome P17 (CYP17). Abiraterone acetate blocks this enzyme, which stops both the testes and other tissues in the body making testosterone.

Researchers are looking at using abiraterone treatment for men with prostate cancer that has spread to another part of the body, and who have had hormone therapy that is no longer working.

Chapter Summary

Plan your presentation with the audience in mind. For decades, scientists primarily communicated only with each other. Today, explaining science to non-scientists is crucial for their support and understanding.

Science communication takes place in three situations:

1.   publications
2.   presentations
3.   media

They all overlap, producing the three overlapping rings.

Read, watch and listen to great science communicators, and aim to emulate them. Make things as simple as possible, but no simpler.

2

The Seven Challenges of Communicating Science

Conveying the risks and benefits – the problem of information overload – the importance of telling a story – less is more – not seeing the wood for the trees – don't ignore the obvious – a tragic misuse of PowerPoint – the difficulty with jargon – clarifying your objective – attitude softening – explaining the benefits – the power of appropriate simplicity – what to include – what to exclude – examples of excellence.

Defining the Difficulties

In the previous chapter I examined why it is important for experts to communicate about science, and gave some examples of where it has been done well. When you see it done well, it seems so easy, like watching a great tennis player or golfer. You think 'Why can't I do that?' or 'How hard can that be?' In reality, we all face many obstacles to communicating clearly. Scientists in particular face specific challenges. Einstein said, 'A problem correctly described is 90 per cent solved.' With that in mind, in this chapter I want to describe the major problems faced by scientists when they aim to communicate clearly.

Balancing the Risks and Benefits

The former US President Harry Truman was so frustrated at the advice he received from his economics advisers ('On the one hand, Mr President … but on the other …') he once said, 'Give me a one-handed economist!' Scientists are notorious for behaving in the same way. Their training encourages them to view any problem from all possible angles, and they are reluctant to come down on one side of an argument until they have all the facts.

In reality, we rarely have the luxury of 'all the facts'. What we have at best is 'all the facts we have been able to establish so far based on the limitations of the trial design'. So we have to make a judgement on an incomplete data set. Even a huge clinical trial of 15,000 patients may not spot an adverse event that occurs once in 20,000. This is where your experience and intuition come in. As the presenter you have a responsibility to make a judgement, based on the data, about whether the overall risk/benefit profile of a new treatment is favourable or not. Your job is not just to *report* the data, but also to *interpret* it for the audience. This task of balancing the positive and negative underpins every example of science communication.

It is hugely important to you at both a personal and professional level. Your personal credibility is at stake if you appear to overstep the mark and sound too positive. You then run the risk of being accused of hyping the data and possibly being in the pocket of the sponsors, usually a large, rich pharmaceutical company. At the same time you want to be as positive as the data allows. You don't want the audience to be put off unnecessarily or prematurely from continuing with what could be an important new medication for patients. Finding the right balance of just the right amount of positivity and concern can be difficult. There is hardly a new drug approved by regulators anywhere in the world where somebody has not put a different interpretation on the trial data.

So how do you find 'The Goldilocks Option' of just the right amount of risks and benefits in your talk? You do this in a number of ways. Primarily by deciding what to include and exclude. My advice here is to follow the process adopted by publicly-quoted companies when deciding whether to inform the stock markets about an important development: if it can be deemed material, you should disclose it. If you have a reservation about a finding, or a concern about a potential problem, it is your responsibility to include it in your talk. That's why you are on the podium and they are in the audience.

The other way you convey the risk/benefit equation is by using the right language to describe a finding. You need to convey the extent to which you find the data convincing. On a scale of robustness, you might say the data:

> *Appears to suggest ... suggests ... supports the idea that ... shows a trend towards ... demonstrates ... proves ...*

By combining these two techniques you can accurately convey your own interpretation of the data. That is your starting point. I now want to turn to the seven challenges of communicating science. There are seven challenges:

1. Information is not communication.
2. 'I know so much, I don't know where to begin'.
3. It's not about the PowerPoint.
4. Great communication means saying something in a way which cannot be misunderstood.
5. Start with the end in mind.
6. Don't confuse features and benefits.
7. Make everything as simple as possible, but no simpler.

Let's look at them in more detail.

Challenge 1: Information is Not Communication

On 30 September 1950 the *British Medical Journal* (*BMJ*) published the first instalment of what would become a lifetime's work for a British epidemiologist called Richard Doll and his colleague Austin Bradford Hill, a statistician. The title was 'Smoking and carcinoma of the lung. Preliminary report'. In it, the pair of researchers highlighted a 'phenomenal increase in the number of deaths attributable to cancer of the lung'.

They quoted a compelling statistic to support their claim: In the quarter century between 1922 and 1947 the annual number of deaths recorded [from lung cancer in the UK] increased from 612 to 9,287, or roughly fifteen-fold. They said, 'This remarkable increase is, of course, out of all proportion to the increase of the population – both in total and in particular in its older age groups.' They discussed how a number of possible causes had been considered and explained how they had set about investigating a link with cigarette smoking. After outlining their research methods, presenting their findings of the incidence of different cancers in smokers and non-smokers, they reached a conclusion of stunning simplicity:

> *To summarise, it is not reasonable in our view, to attribute the results to any special selection of cases or bias in the recording. In other words it must be concluded that there is a real association between carcinoma of the lung and cigarette smoking. We therefore conclude that smoking*

is a factor, and an important factor, in the production of carcinoma of the lung.

Over the next 50 years, Doll and Bradford Hill published many more papers on the health consequences of smoking. Much of their research involved the smoking habits of male British doctors ... an impressive 34,439 took part over the 50 years from 1951 to 2001. The *BMJ* published an update in 2004 under the title: 'Mortality in Relation to Smoking: 50 years' observations on male British doctors'.[1]

Doll and Bradford Hill, supported by other colleagues including Richard Peto, conducted ground-breaking research which was soon complemented and supported by many other researchers around the world. It quickly became apparent that smoking caused not only lung cancer, but could exacerbate many other diseases too.

In 1954 Doll and Bradford Hill published a paper confirming the link between smoking and lung cancer, and three years later the British Medical Research Council announced there was 'a direct causal connection'. In 1962 the Royal College of Physicians concluded that smoking is a cause of lung cancer and bronchitis, and said it probably contributes to heart disease. Three years later, cigarette advertising was banned on British TV, and health warnings on cigarette packets were introduced in 1971.

From 1950 the evidence that smoking is bad for your health began to accumulate, and received large amounts of publicity. During that time, people were still smoking in the face of clear evidence of the harm it could cause. One of them was my father. Like many people of his generation he started smoking when he was about 13 years old. Coincidentally, this was about 1950, the time that Doll's first smoking research was published. As Doll and colleagues continued to publish, and the evidence mounted, my father, and millions like him, continued to smoke.

I have always been an anti-smoker, from the days when, as a small child, I was forced to sit in smoke-filled rooms with my parents. I have lost count of the times I told my father to quit. I quoted the research, told him it was worth quitting however old you are, and banned him from smoking in the house.

---

1 You can read the original 1950 paper here: http://www.ncbi.nlm.nih.gov/pmc/articles/ PMC2038856/?page=12004 ; or the 2004 paper here: http://www.bmj.com/content/328/7455/1519. full.pdf+html)

Nothing worked. Then one day, when I was a TV journalist, I interviewed clinical pharmacologist Professor Peter Sever at St Mary's Hospital in London. He had been conducting research into the effects of smoking on cardiovascular risk, in particular on the speed at which cardiovascular (CV) risk is reduced when you quit.

In my TV interview he said this:

> *We know that smokers have sticky blood. The viscosity, or stickiness, of a smoker's blood is thicker than a non-smoker's. That's one factor that puts them at risk of a heart attack or stroke. What we've done in our research is to measure the viscosity of the blood in people who are on stop-smoking courses. We've measured it up to their quit day, and continued to measure it afterwards. We've discovered that their blood is measurably thinner within 48 to 72 hours of their last cigarette. To put it another way, if you stop smoking on a Tuesday, you start to cut your risk of a heart attack or stroke by Thursday or Friday.*

As soon as Professor Sever said this, I knew this would make a great piece for the evening news. When I next saw my father I showed him a video of the interview and an amazing thing happened: He stopped smoking after nearly 50 years. All the research conducted by Doll then by Richard Peto and others meant nothing to him. A single television interview convinced him. There was no lack of information that smoking is bad for you. As far as my father was concerned, it hadn't been communicated in the right way.

The story nicely sums up the first challenge of communicating science:

> *Information is not communication.*

I was recently engaged by a large pharmaceutical company to help them prepare for a regulatory hearing for a new drug. Their submission contained 802 PowerPoint slides! Of course, they were not planning to present them all, and most were in the back-up set. However, it was obvious that their case was completely unclear and unfocused. In a recent US lawsuit involving another large pharmaceutical company, it emerged the company had provided 14 million pages of documents to plaintiffs' counsel! This of course may have been a case of deliberate obfuscation, but if so it still illustrates the point.

This is a classic problem for scientists. In their desire to appear thorough, they equate information overload with communication effectiveness. This leads to a belief that 'More is better.' In reality, the opposite is true.

In the scientific world, information is valued higher than almost any other property. In the world of research, data is prized above all else. It is understandable that the inhabitants of that world, having collected data or gathered information, regard their work as done. However, if you regard the onward communication of that information as a key part of your role, you have more work to do.

Here are some ways to ensure you focus on communication, not just information:

- 'just the facts' is not enough
- in science, information is the platform for your talk, interview or presentation. Without it, you would have nothing to say. But information alone is not enough.
- tell a story and build a compelling case
- establish relevance
- stress benefits
- make emotional connection
- anticipate questions

Challenge 2: 'I know so much, I don't know where to begin'

I regularly use this cartoon when I make presentations about the challenges of communicating about science:

It always receives a wry smile, because many of the people in the room know it applies to them. At the same time, being accused of knowing too much is flattering … it takes a really clever person to know *too* much about something, right? That's far better than not knowing enough, surely? Oh how superior it makes us feel!

The cartoon neatly summarises a key challenge for academics and scientists. In your world, knowledge is the basis of your success, and it is impossible to know too much. Part of the reason the cartoon gets a laugh is that the very idea of 'knowing too much' is itself laughable. How can anyone know too much?

"I know so much I don't know where to begin!"

Figure 2.1   'I know so much I don't know where to begin!'
Raymond Patmore

The problem of knowing too much is that it's difficult to stand back and decide what is important. Many languages have their own version of the English idiom, 'You can't see the wood for the trees', or as Americans and Australians say 'You can't see the forest for the trees.' It refers to someone who is unable to understand what is important in a situation because they are paying too much attention to details. This is a classic problem: After you've spent years researching a single topic you get to a point where you can't see the wood for the trees.

Some years ago I was working with a pharmaceutical company whose best-selling drug for many years had been a beta blocker, a type of drug known as an anti-hypertensive, used for reducing blood pressure. A rival company had included a similar drug, another beta blocker, as the comparator arm in a study of their own new and exciting drug, which had a different mechanism of action. The rival company hoped that the trial would demonstrate that their exciting new kid on the block drug would be safer, more effective or both. They would then say there was no longer any need for beta blockers, as they had been superseded. My clients expected the beta blocker would come off second best against the newcomer, but felt there was still a need for it in specific patient groups. They asked me to help them craft a story which explained this, and helped to defend their product's position after years of success and millions of patients successfully treated.

I ran several meetings with senior people in the company, trying to help them tell the story clearly and fairly, without making any claims that could not be substantiated. As usual, I was aiming for the combination of the accuracy of peer-reviewed science and the narrative skills of journalism. At the end of every meeting, we all felt we had made progress. First we developed a story line, then key factual statements, then different versions for scientific and non-scientific audiences, from journalists to payers and patient groups. Finally we agreed the question and answer documents (general and scientific), about the meaning of the trial and future prospects for our beta blocker.

Any statements a pharmaceutical company makes about its medicines need to comply with strict, legally binding codes. So all our statements need to be approved internally by a number of people with different responsibilities, including legal, regulatory, medical and investor relations. On this occasion everyone was happy except for the person who had to sign off statements regarding the mechanism of action of this and similar types of drugs. He was an associate professor at one of the world's oldest universities, and was renowned as one of the world's leading experts on beta blockers. He was also very busy, and never had time to meet us personally. We sent him our ideas, and he sent back comments, which were usually negative and technical.

On several occasions, he crossed out key sections of our argumentation, with no explanation, just the comment, 'You can't say that' or 'This is incorrect.' Nobody in the team I was working with understood exactly what was wrong with our explanation, or why we couldn't say it. Increasingly frustrated, and with time running out before the major conference at where the trial results were due to be announced, we finally sat down with the professor, face to face.

He had worked on beta blockers for 28 years. He was a charming man, and his knowledge was truly impressive. He had so much gravitas and obvious learning in the way he explained things, and such a specific way of speaking that nobody would disbelieve a word he said. Unfortunately he suffered from a serious case of knowing too much.

A side effect of knowing too much is often that you don't realise how little others do know. This is often because something is so obvious to you that you wouldn't think of stating it any more than you would say, 'The sun will rise in the East tomorrow.' It would be a statement of the obvious. This was the case here. The problem often manifests itself to the audience when something doesn't make sense. You feel there's a missing piece of the puzzle which nobody has given to you. You go round it again and again, and keep feeling something is not right. One problem is that many people are afraid to ask the questions that might provide the missing piece, in case it makes them look foolish. As a journalist, I have years of experience of asking questions that might make me look foolish, so it doesn't worry me. I questioned him closely, asking what he regarded as more and more inane questions, and quite quickly we got to the bottom of it.

The missing piece of the puzzle was that not all beta blockers are alike. In particular, some are hydrophilic, while others are lipophilic, and lipophilic ones tend to cause more side effects. All our statements about beta blockers were only true for some of them. We needed to divide the type of drug into two, right at the start of our explanation. When I said to the expert, 'So our drug is different from the drug in the study because ours is hydrophilic, while theirs is lipophilic?', he laughed and said, 'Yes, of course. Maybe you should brush up on your pharmacology, John!' Laughter broke out around the room and I sensed most of the tension draining away. Now we had a story we could tell!

Contrast that story with another one where I was again involved. The leading expert here was an eminent British cardiologist, Professor Peter Sleight. At the time of writing (2011) he is Emeritus Professor of Cardiovascular Medicine at the University of Oxford. He was responsible for setting up the first ever large-scale, multi-centre trials on cardiovascular drugs, which led to radical changes in the way drug trials are conducted in all areas of medicine around the world. In 2010 he received the Gold Medal of the European Society of Cardiology (ESC) at the congress in Stockholm. In addition to his superstar status, he is a wonderfully clear communicator. Unlike many, knowing so much never means he doesn't know where to begin.

I hosted a press conference at the ESC in Munich where Professor Sleight was the main speaker. The subject was a follow-up of the trial I mentioned in the previous chapter, comparing two different types of cardiovascular drugs, an older one called an angiotensin converting enzyme (ACE) inhibitor with a newer one called an angiotensin receptor blocker, known as an ARB. The audience were medical journalists, and after a few introductory remarks from me, I gave the floor to Professor Sleight. His opening was a model of scene-setting which I encourage you to follow when you present medical data:

When I have finished a trial I have often forgotten what we were studying at the beginning, so for journalists who don't do this kind of thing every day it must be worse. So I will first remind everyone what it was about. It was a straight comparison between the 'new kid on the block' an ARB, in this case what looks like the best of the ARBs in many ways, longest acting, fewer side effects and so forth, and a very good ACE inhibitor, at full dose, versus the combination, ie the two drugs together, again at full dose.

On this slide you can see the three arms ... ARB v ACE v combination. In particular, notice that the ACE inhibitor was used at full dose, both on its own, and as part of the combination. That's different from other trials where you've had a combination of ACE and ARB, because very often the ACE dose has been reduced, which of course makes the ARB look a little better. Here, though, we had a full dose of the ACE so that's a hard test. What we wanted to find out was whether one drug is a better choice than the other, and whether the two together would be even better....

What's so good about this is that the professor sets out his stall right at the start of the presentation, demonstrating that however much he knows about his subject, he certainly knows where to begin.

Where to End?

The professor in my cartoon says he doesn't know where to start. He has another problem: He doesn't know where to stop. We have a phrase in English, 'A little knowledge is a dangerous thing.' As another professor said to me, 'But sometimes a little knowledge is all you need.' Einstein (again) put it better: 'A little knowledge is a dangerous thing. So is a lot.'

Knowing where to stop is as important as deciding where to start. When people ask me 'How long should a presentation be?' I usually reply, 'As short as possible, but as long as necessary.' The key is to prioritise what information you want to communicate:

- What is essential for the audience to know?
- What would be helpful for them to know, to help their understanding?
- What else could you tell them if you had enough time and they had enough interest?

Start with the essentials. Then if your allotted time slot is long enough, go down the list. You are searching for what we call The Goldilocks Option ... not too much information, not too little information, but just the right amount of information. We will explore this in detail later in the book, and look at some techniques for achieving it.

Challenge 3: It's Not About the PowerPoint

When I make this point in my own talks about effective presenting, almost everybody nods and smiles. Occasionally, somebody cheers (or whoops, if we're in the US). The problem is that in many presentations, PowerPoint becomes a replacement for the speaker, instead of a re-enforcement. When this happens the speaker is just supplying the audio track for the slides, which takes away many of the presenter's strengths. In particular, it strips them of their ethos, their reputation and standing with the audience. (See Chapter 5 and Aristotle's methods of persuasion for more on this.)

Considering it's just a piece of software, the hatred that PowerPoint engenders is truly remarkable. You would be forgiven for thinking that using it is compulsory. Look it up on the internet and you will find articles with titles like 'PowerPoint is evil' and 'Is PowerPoint the devil?' The Dilbert cartoons about office life regularly parody the use of it, and Dilbert himself has referred to his audience falling into 'A PowerPoint coma'. The phrase 'Death by PowerPoint' is widely used in the business and academic worlds.

In Switzerland a man set up the Anti-PowerPoint party and expected to win enough votes in the election to win a seat in Parliament. PowerPoint was even blamed for the Columbia Space Shuttle disaster in 2003, when the spacecraft disintegrated over Texas while re-entering the earth's atmosphere. The tragic sequence of events began when a piece of foam insulation broke off on take-off. Engineers urgently attempted to predict whether the missing piece might cause a serious problem on re-entry. They made their report to senior managers, while Columbia was still in space, in a PowerPoint presentation of 28 slides, rather than in a detailed engineering report. As the two inquiry reports said:

> *Many of the engineering packages brought before formal control boards were documented only in PowerPoint presentations. It appears that many young engineers do not understand the need for, or know how to*

prepare, formal engineering documents such as reports, white papers or analyses.[2]

Criticism focused on one complicated slide in particular:

As information gets passed up an organization hierarchy, from people who do analysis to mid-level managers to high-level leadership, key explanations and supporting information is filtered out. In this context, it is easy to understand how a senior manager might read this PowerPoint slide and not realize that it addresses a life-threatening situation.[3]

Medical analogies abound among the anti-PowerPoint forces:

Imagine a widely used and expensive prescription drug that promised to make us beautiful but didn't. Instead the drug had frequent, serious side effects: It induced stupidity, turned everyone into bores, wasted time, and degraded the quality and credibility of communication. These side effects would rightly lead to a worldwide product recall.

In my view, blaming PowerPoint for long, boring presentations with no clear focus and small, unreadable text is like blaming cars for drunken drivers. It's shooting the messenger. PowerPoint is innocent! It's the users who are guilty! PowerPoint is merely a tool. What you do with it is a matter for you. To turn to one more medical analogy, I agree with another writer who compared the abuse of PowerPoint to the abuse of antibiotics. When penicillin was invented it was regarded as a silver bullet for a wide range of illnesses affecting humans and, later, animals. It was used with such enthusiasm that its original purpose was lost. Over time the use became so prevalent that resistance developed and the antibiotics became less and less effective. So it is with PowerPoint, but that's not the fault of the people who developed it.

What this discussion illustrates nicely is that your presentation is not about the PowerPoint. If all you want to do is to read the slides, just send the slides, and don't bother to turn up yourself. When I'm invited to speak at a big meeting or conference, the client often asks to see my slides in advance. Usually they

---

2   From the Return to Flight Task Group, set up after the disaster.
3   From the report of the Columbia Accident Investigation Board.

need to go through legal and regulatory approval. I have no objection to that. However, if they ask to see my 'presentation' I tell them that the presentation is the combination of me and the slides. I am an important part of it! You should do the same ... it's as much about you, and what you say, as it is about the slides. The data, or the information, is the bedrock of your presentation, but it really is 'not about the PowerPoint'. We will look in detail at designing and using PowerPoint slides in a later chapter. I will also ask you to consider whether you always need PowerPoint (or any other presentation software) at all. So if your presentation is not about the PowerPoint, what is it about? Here are some suggestions:

- telling a story
- explaining the data
- building on the information you have
- putting the findings into perspective
- explaining the relevance to the audience
- generating enthusiasm for your story
- adding to what is on the slides
- producing a memorable experience

Challenge 4: Great Communication Means Saying Something in a Way Which Cannot Be Misunderstood

This is my mantra, and is printed on much of the material my company produces. The full quote is:

Good communication starts with saying something in a way which can be understood.

Great communication starts with saying something in a way which cannot be misunderstood.

As I travel around the scientific, medical and pharmaceutical worlds, I meet a lot of good communicators who meet the first challenge here, that is, what they say can be understood. However, the really good ones also meet the second challenge, and cannot be misunderstood. The first point to notice here is the phrase 'begins with ...' This is because successful communication involves taking into account the way the audience is receiving what you are telling them, then adapting as necessary. For example, if they don't appear to understand the technicalities, make it simpler; if they don't get the significance of the data to their work, explain it differently; if they're not convinced about the 'big picture' of your hypothesis, give them a real-life illustration.

Communication didn't always include a feedback loop. It was only when Shannon and Weaver published a book titled *The Mathematical Theory of Communication* in 1949 that feedback was generally recognised as an important element of communication. See Chapter 3 for more on this.

Here is another cartoon I use in presentations, to illustrate the importance of using language which cannot be misunderstood:

Figure 2.2   Example of miscommunication
Oliver Preston

So how do we ensure we cannot be misunderstood? By presenting understandable concepts, expressed in the right language.

An understandable concept is one which the audience can grasp. If you want to introduce a non-medically qualified patient to anti-clotting medication you would not talk about the coagulation cascade and whether it was more effective to block it by inhibiting Factor Xa or thrombin, and the importance of the von Willebrand factor. The concept would be meaningless, and the patient likely to end up more confused than when they started. You would be better advised to keep it simple.

The right language is a key element of successful communication, and causes a real problem for medics and scientists. Medical students learn a new language of 6,000 words at medical school. Non-medics don't speak that language, so you need to translate it for them.

The problem is actually more complicated than that, as there are in fact two types of jargon:

1.   Words and phrases which have different meanings to specialists and non-specialists.
2.   Words and phrases which non-specialists do not understand.

An example of the first type is the word 'censored' as used by statisticians. Critics of the way clinical trials are conducted say 'they even admit themselves that they censored the data', suggesting the data has been twisted in some way, to fit the conclusion the sponsors wanted (usually about a new drug). In reality, 'censored data' in this sense means data which is incomplete.

A medical example is the word sinus. To lay people, your sinuses are in your nose. To a cardiologist, sinus rhythm is a term used to describe the normal beating of the heart, as measured by an electrocardiogram (ECG). Classic cases are the words 'chronic', which to lay people means 'bad' or 'serious'.

Acronyms and abbreviations can also cause confusion, and I am not always innocent in this. I recently tweeted that I was on my way to Lugano, Switzerland, to host a conference on NHL, meaning Non-Hodgkins Lymphoma. 'I didn't know you were working with the National Hockey League', a friend replied. NHL is a good example of what we call TLAs … Three Letter Acronyms. Avoid them. A more serious example is the word 'progression' when used by oncologists.

If your oncologist tells you that your cancer is progressing or 'has progressed' you would be forgiven for thinking that this is good news. 'The recovery is progressing well', is good news. 'The cancer has progressed' is not, as seen in the popular endpoint in trials of cancer drugs, Time To Progression.

Doctors and scientists are traditionally major culprits of the second type of jargon, using words and phrases that mean nothing to anyone else. It is understandable because at medical school and with colleagues you are expected to use the medical and technical terminology. When I give talks I introduce this section with an old photograph of a distinguished-looking, grey haired male doctor and a British schoolboy, in the consulting room. The doctor is saying, 'You've got a bad case of acute diverticulitis' and the schoolboy says, 'but what about my dodgy tummy?'

In recent times I have seen two examples of communication which went wrong for very different reasons. The first was at an oncology conference in Switzerland. I was working with the Principal Investigator (PI) of a major trial of a new type of drug used to treat breast cancer. The trial's finding was very important, and has been one of a number of studies which have now changed clinical practice. The professor was due to present the findings to the conference the following day, and be interviewed by a medical journalist from The Wall Street Journal, the distinguished and very reputable US newspaper renowned for checking every detail of its stories. I could see that although the PI was a leading scientific researcher, communication was not his strong suit. He was the walking, breathing example of the professor who knows so much he doesn't know where to begin. He was also doing the presentation and interview in English, his third language.

I stressed the importance of using the right kind of language, told him it was vital to keep it simple and be very clear about the findings. I advised him to listen carefully to the journalist's questions, as they would give us a pointer regarding her level of understanding. When the journalist came on the phone, the professor ignored all that and went straight into a jargon-filled monologue which anyone but an experienced cancer researcher would have had difficulty following. The journalist did her best to keep up, but it was a losing battle. When it was over, I said, 'Well, professor. How did you feel that went?'

He replied, 'It was just ridiculous. She didn't understand any of it. She wasn't even a biologist!' Leaving aside the scandalous implication that biologists are the lowest form of scientific life, the professor had completely missed the point, which was that his responsibility was to communicate the findings of the trial, and their implications. Fortunately someone else on the team with better communication skills was able to call the journalist back, and explain it more clearly.

The other example was completely different, but by coincidence it was at another oncology conference. I was speaking at a satellite meeting with a British oncologist, a specialist in kidney cancer, a disease for which at the time of writing there is no cure. I do sometimes have the privilege of meeting excellent communicators, who have a natural talent for explaining complex things very clearly, and have worked hard at it. I find them a joy to listen to. This doctor was one of them. He told a story of how he had recently explained to a patient that he had early stage cancer, and explained the treatment options. Having met the oncologist, and seen how clearly he communicates, I am sure he explained it all very clearly. At the end of it he asked the patient 'Is there anything that is not clear, or anything you want to ask me?'

The patient replied, 'Well I'm just glad I haven't got a tumour, doctor, because my mother died of a tumour.' The doctor had avoided using the word 'tumour' because it risked confusing the patient. (Some tumours are cancer, some are not.) However, in doing so, he had inadvertently made the patient's condition seem less serious than it was, and hadn't communicated clearly at all. I thought it was interesting that the doctor chose to tell me this, and tell the story against himself. It showed how keen he was to be seen as a clear communicator. Of course there are many other factors involved in dictating what patients hear when their doctor gives them a diagnosis, but the first step is to use the right language. This story shows how careful you have to be.

It may be that everybody in every audience to whom you present medical or scientific data knows the difference between a tumour and cancer. However, the lesson is clear: It is your responsibility to ensure that what you say cannot be misunderstood.

Here are some words which can cause confusion to non-specialists, so should be used with care:

incidence	prevalence
benign	malignant
morbidity	mortality
chronic	acute

The presentation of scientific data is an area which is ripe for confusion and misunderstanding. I have lost count of the number of physicians who have said to me, 'I hated statistics at college, and I've never got over that feeling. I get the general idea, but I can't get to grips with the detail. I just accept what the statisticians tell me.' I can understand that, and within large pharmaceutical companies it often occurs to me that the only two types of specialists whose views cannot be challenged are statisticians and lawyers. However, to present data effectively, you need a grasp of statistics, and an understanding of basic statistical terms. Then, and only then, can you start to explain them, and their importance, to your audience. They don't need to know the details, just the significance (itself a specific statistical term!). I once hosted a seminar for journalists about clinical trial design and statistics in medicine. To understand their knowledge level at the start, I asked them a number of questions. One of them was, 'What do you understand by the expression, "p-value"?'

One senior journalist answered, 'I don't understand how it's calculated, but I look for lots of zeros after the decimal point. The more zeros, the more reliable it is.' I thought that showed a good understanding overall, even though it ignored an important fact. This is that a p-value of 0.05 or less means that the finding is statistically significant. That means the likelihood of the result being down to chance is less than 5 per cent, which is the arbitrary level at which we say we're confident it's true. However, she had grasped the overall importance.

We will address this question in more detail in a later chapter, about interacting with your slides.

Challenge 5: Start With the End in Mind

I sit through hundreds of scientific presentations a year, all over the world. Occasionally, somebody I know slips into the seat next to me, just as the talk is ending. Sometimes they say, 'What was the point of that?' Sadly, I often have difficulty answering the question. Random thoughts run through my mind: 'That was a big study' … 'It had a long follow up' … 'He said it was important' … 'I wasn't sure what was new and what was already known …' 'I'm not sure what it means to clinical practice …' 'I didn't feel the conclusion was sufficiently supported by the evidence, but maybe I was confused …'

This can be avoided by adopting the 'Start with the end in mind' approach. It's one of 'The 7 Habits of Highly Effective People', an international best-selling book written by American lifestyle and management guru Stephen R. Covey. (His phrase is actually 'Begin with the end in mind.') The book has sold 15 million copies in 38 languages since it was first issued in 1989. The tip has particular relevance for scientific presentations and conversations.

When I work with presenters on important presentations, I ask them to ask themselves two key questions before we start work on the talk:

What do I want to achieve by this talk?

What do I want the audience to do, think or feel after they have heard me?

Of course the answers are a matter for you … it's your data, your story, and you have the clearest idea of what you want to achieve. Here are some possibilities:

- I want the audience to realise that there is new treatment option in X disease which is safer and more tolerable than the older drugs for many patients, without paying the price of reduced efficacy.
- I want them to invest in my idea.
- I want them to realise that this could be really important, and we need to fund bigger studies in a specific patient population.
- I want them to employ me.
- I want them to understand why the newer drugs should be used on top of the old ones, not instead of them.
- I want them to realise the significance of early diagnosis of condition X.
- I want them to understand what side effects to expect in a small number of patients, how to recognise them and what to do about it.
- I want them to be aware of the prognostic/diagnostic indicators which we can now use.
- I want them to be aware that a genetic test can identify likely responders with 95 per cent accuracy.

The key point about all these objectives is that they are very clear. The next time you plan a talk, I urge you to produce equally clear objectives. Once you have answered the questions about your aims and what you want them to do, you need to move on to the next list of questions:

- I know what I want them to do … now what do I need to do to achieve that?

- What do I need to tell them?
- How much context do I need to include?
- What will prevent me achieving my aims?
- What is the strongest data I have that will convince them?
- What objections will the audience have, and how can I overcome them?
- How can I best illustrate my talk?

Having done that, you can now start to sketch out your talk. I will outline some techniques for doing this in the next chapter.

Overcoming Objections

One key question you need to address concerns potential obstacles to you achieving your aims. In particular, you need to address concerns and objections which the audience have. A useful technique here is known as *attitude softening*. It's based on a psychological concept called reciprocity. This means that people are socially programmed to behave in the same way towards you as you do towards them. If you smile at someone they almost always smile back. If you are objectionable, the temptation is to do the same. An attitude softener works the same way and is widely used in selling: You give credit for their wise thoughts, then express their objection as a need, which you can then meet. In presentations we use it less to get what we want, but more to defuse a potential objection before it is expressed.

Let's look at an example. I was working with a physician who had been involved in a big trial on the first oral medication for Relapsing Remitting Multiple Sclerosis (RRMS). One of the key treatment aims in this area is to reduce the number of relapses experienced by the patient in a year. The most common treatments for RRMS are a group of drugs known as the beta-interferons. He was presenting the key data at an international conference.

When he reached the section about reducing relapses, he put up the data, comparing the new drug with the beta-interferons, and said, 'You can see here that the new medication was more successful at reducing relapses than the beta-interferons. When I've shown this to people they have sometimes said the difference is not very big. However, what is important to realise here is that these patients in the control group are on beta-interferons, which themselves are very good drugs, so any improvement on that is really worthwhile.

That's what we saw here ... and improvement on something that's already regarded as being pretty good. The difference is statistically significant, but also clinically important to patients.'

There are several benefits to the attitude softening technique. One is based on a concept called *primary definers*. They are the people who define an issue first, in a particular way. Pressure groups and lobbyists are often very successful at this. For example, anti-vaccine campaigners who were against the combined Measles, Mumps, Rubella (MMR) vaccine claimed that giving the three vaccines at the same time was unsafe. They claimed it could cause autism and Crohn's Disease in young children. In communication terms, they were the primary definers, and defined MMR vaccination as a safety issue. In contrast, governments defined it as a public health issue. They claimed that if parents had to bring their babies to the surgery two or three times for separate injections, many of them would not do so, vaccination rates would drop, herd immunity would not be achieved, and there would be outbreaks of the three illnesses. To achieve the public health objectives, the governments needed to first overcome the safety concerns expressed by the primary definers. In some countries, notably the UK, this took a lot of time and money.

The cost of new medications is frequently a difficult issue, but in my view one which needs addressing head-on. In your talk about a new drug, you can define the cost as a moral issue, that is. 'How much is better quality of life worth for our patients?' Or you can accept that the cost is an economic issue, but define it more broadly. Here you might say, 'It is true that the new drug is more expensive than the older ones. However, with the better safety and tolerability profile, we expect better compliance. With the old drugs, 30 per cent of patients stop taking them within six weeks because of the side effects. As a consequence, they need more consultations with the physician and nurse, are tried on other drugs which often have different side effects, and they end up in spiral. Every twist of that spiral costs money. Our view is that with the better compliance we expect, most of these extra costs can be avoided.' Of course, you have to have the data to back up the claims!

Another benefit of the attitude softening technique is that you are preparing the ground for a discussion about the objection. You are signalling to the audience that you understand their concerns, have thought about them and have a solution to propose. As with any communication exercise, seek first to understand, then to be understood. Attitude softeners help you achieve this. I will return to this topic later, in the section on handling Q&A sessions.

Challenge 6: Don't Confuse Features and Benefits

One of the first things anyone asks themselves when attending any kind of talk is the question, 'What's In It For Me?' (Often abbreviated by Americans to WIFFM?). You need to answer this question clearly. If this was a marketing book I would say the answer to WIIFM is your value proposition. It's whatever you are selling that's worth something to your potential customers. In the scientific world we are not selling anything, at least not in the sense of a commercial transaction where money changes hands. However, please don't think that just because your audience are not paying hard cash there is no exchange involved. You are selling them an intangible. It may be your idea, concept, data, experience or expertise. They are paying for it with their time and attention. And they expect value for this just as they expect value for money in a commercial transaction.

So how do we deliver value to an audience? First, by understanding what they are looking for, as spelt out in the discussion of the earlier challenges. Then we come to the crux of challenge number six: Don't confuse features and benefits. This is another well-known sales or marketing technique which you need to adapt for the scientific arena. To answer the question, 'What's in it for me?' You need to spell out the benefits. The problem is that a lot of scientific research starts with the features. As an example, the way a drug works, or interacts with other drugs, is its feature. The benefit is what that means.

The relationship between the two is based on this idea:

Feature … which means that … Benefit

It's the phase *which means that* which translates the feature into the benefit. To grasp the concept before we apply it to scientific talks, here are some examples:

Feature: My phone has a 50 number speed dial facility.

Which means that …

Benefit: I can call colleagues and friends really quickly.

Feature: My car has a driver's airbag.

Which means that …

Benefit: If I'm in a crash I'm less likely to be injured by my head crashing into the windscreen.

Feature: My computer has a remote mouse and slide advancer.

Which means that …

Benefit: I can move around the room during presentations, and really engage with the audience rather than being stuck behind a podium.

Now let's translate the concept into medical and scientific presentations:

Feature: This new drug is more specific to X receptor.

Which means that …

Benefit: It has fewer side effects than the old drugs.

To a different audience, you might take the story one step further:

Feature: This new drug has fewer side effects than the old ones.

Which means that …

Benefit: Patients continue to take their medication, so compliance is improved.

Feature: This new anti-platelet agent prevents platelets clumping together, but without increasing the risk of internal bleeding.

Which means that …

Benefit: Your risk of a stroke or heart attack caused by a blood clot is reduced, without dangerous side effects.

Challenge 7: Make Everything as Simple as Possible, but no Simpler

This is one of many quotes attributed to Einstein. It now appears that he didn't actually write this exact phrase in any traceable document, though he certainly agreed with the concept and of course he could have said it verbally. Reducing complex matters to the simplest level possible was constant theme of his writings. He did say:

> *If you can't explain it simply, you don't understand it well enough.*

He also apparently said:

> *It should be possible to explain the laws of physics to a barmaid.*

The quote which comes closest to the one which concerns us here is:

> *It can scarcely be denied that the supreme goal of all theory is to make the irreducible basic elements as simple and as few as possible without having to surrender the adequate representation of a single datum of experience.*

So whatever he actually said, his point was the same: Make everything as simple as possible, but no simpler. This is our final challenge, and it is crucial to recognise that both parts of the advice are equally important. If you don't make it as simple as possible it's still too complicated and you leave yourself open to the problem of Challenge 4, that is, you can be misunderstood. If you stretch the simplicity too far, you may lose the scientific validity of whatever it is you are explaining. However, including everything can make things more complicated, rather than simpler. You need to strike the right balance. Once again we are trying to establish 'The Goldilocks Option' of just the right amount of simplicity.

So where is The Goldilocks Option? That depends on the subject matter and the audience. Let's consider the presentation of the results of a clinical trial comparing Drug X with Drug Y for the reduction of risk of event A in patients with condition B. The trial will reach a result with a specific patient population at a specific time. Based on a number of factors, including the trial design, relevance of the patient population to clinical practice, the dosing regimens used and statistical methods, it is reasonable to extrapolate that result to similar populations. However, both the trial result and the extrapolation are full of caveats, qualifying factors and nuances.

> *Not making it simple enough would mean leaving them all in. Making it too simple would mean taking them all out.*

It's your job to decide where the balance lies.

Let's start with the trial design, the first crucial aspect of any study. When doctors and researchers present the design, they usually begin with the inclusion and exclusion criteria. This often involves reading a list of gender, age, duration of disease, previous treatment, blood counts, tumour grades or liver function test (LFT) results where appropriate and other factors such as comorbidities, weight and smoking history. You would then describe the methodology, including how patients were recruited, randomised and treated. This would include detail of the drug regimens in each arm of the study. This is generally too much detail for you to enumerate in a presentation, though it needs to be in the written publication. (There are rare exceptions, which I will discuss later in this section.) You may even feel that you need to include it all on a slide for the sake of completeness, but you don't usually need to read it all out. Tell us what's important, and point out what we should take from the criteria presented. There are two important questions here:

- How well-matched are the groups on Drugs X and Y?
- How relevant is this patient population to those seen in clinical practice?

These are the two factors you need to focus on. You might say, 'As you see, the two groups were well-matched in terms of all the important criteria. There were slightly more women than men, but that reflects the incidence of this disease. You'll also notice that the American patients were on average 5kg heavier than the Europeans ... again, that reflects what tends to happen in practice.'

Then you may highlight the key exclusion criteria. 'Patients who'd previously been treated with XXX class of drug were excluded because there is some evidence or cross-resistance, which might reduce the efficacy of drugs X and Y.'

Now let's turn to the other end of the presentation: The results.

The main finding might be that drug X reduced the risk of event A by 14 per cent more than drug Y, with a p-value of 0.003 and a 95 per cent confidence interval (CI) of 12.92 to 16.24. Based on the p-value and the confidence interval, it's fair to say that this is a robust result. In this case we would be justified in including the p-value and CI values in our presentation, but only making a passing reference to them. Assuming you were talking to an informed audience, you might say, 'As you see here, based on the p-value and the confidence interval, this result is pretty robust.'

If you were talking to a less informed audience but still wanted to make the point that these results are robust, you might say, 'These numbers here are calculated using statistical tools which tell us how confident we can be that these results are real, and didn't just happen by chance. The numbers themselves in this instance ... 0.0003 and this range here, from 12.92 to 16.24, tell us these results are very likely indeed to be genuine findings.'

But what if the p-value was 0.05, and the CI was 7.9 to 19.7? That's a far less robust result. Then you would want to draw attention to it, and explain why this is on the edge of what we call statistical significance. You would then have to include more caveats and nuances. If you didn't you would have simplified it too much, and no longer be within range of The Goldilocks Option.

Returning to the subject of how much detail to give regarding the trial design: If the design is in any way unusual (positively or negatively), or is deficient in some other way, you would want to spend more time on it. I worked with one client recently whose major trial, on which they were to base their regulatory submission, included a broad range of patients admitted to cardiac Emergency Rooms (ERs) in the US, whereas the competitor trial had excluded a major sub-set of patients. The details of the trial designs were important, and formed a small but crucial element of the submission.

I worked with another pharmaceutical company whose main competitor had conducted a trial of the competitor's drug versus my clients' drug. The results apparently showed that the competitor drug was better. It was only when you looked closely that it became apparent that the loading dose of my clients'drug used in the trial was lower than that generally used in practice. The discrepancy, in my clients' view, could account for the difference in results. On this occasion, my clients' version of the presentation of the results highlighted this key difference. To return to the point of this section, the competitor's version had over-simplified the situation by ignoring the loading dose question, where my clients felt that this was key.

When I am working with medics and scientists, they often criticise the way journalists summarise the results of clinical trials. I will discuss this more in the chapters on the media, but the main problem is that journalism reduces most things to a binary level of stop/go, black/white, yes/no. In effect, the nuances and caveats are lost in this process and the results or design are oversimplified. So although journalists are excellent communicators, they don't always meet the demands of Challenge 7. That's why journalism and peer-reviewed science don't always go well together.

However, my aim with this book is to make them go together where appropriate, and to encourage you to combine the accuracy of peer-reviewed science with the narrative skills of the journalist. Following Challenge 7 will help you do that. If you are planning a presentation or interview to a non-scientific audience, take a look at some news stories online to get a flavour of how journalists describe your own topic. Look at the language they use, and the level of simplicity versus complexity. Then ask yourself if the journalist concerned has made it as simple as possible, but no simpler. See the later chapters on using the media to communicate science for good and bad examples.

There are many other places on the web to find excellent examples where scientists and medics have met Challenge 7. I mentioned patient websites earlier in this chapter, and will return to them now for more examples. Here is a section from http://www.WebMD.com, a website written by doctors aimed at patients with no medical knowledge:

Leukemia – Topic Overview

WHAT IS LEUKEMIA?

Leukemia is cancer of the blood cells. It starts in the bone marrow, the soft tissue inside most bones. Bone marrow is where blood cells are made.

When you are healthy, your bone marrow makes:

- **White blood cells**, which help your body fight infection.
- **Red blood cells**, which carry oxygen to all parts of your body.
- **Platelets**, which help your **blood clot**.

When you have **leukemia**, the bone marrow starts to make a lot of abnormal white **blood** cells, called leukemia cells. They don't do the work of normal white blood cells, they grow faster than normal cells, and they don't stop growing when they should.

Over time, leukemia cells can crowd out the normal blood cells. This can lead to serious problems such as **anemia**, bleeding, and infections. Leukemia cells can also spread to **the lymph nodes** or other organs and cause swelling or pain.[4]

Macmillan Cancer Support is one of the leading cancer charities in the UK. Their slogan is 'We make things clearer.' Their website is a shining example of The Goldilocks Option of information. Here is an excerpt from their website, explaining how hormonal therapies for breast cancer work.

Hormonal therapies for breast cancer

Hormonal therapies are treatments to reduce the levels of hormones in the body or block their effects on cancer cells. They are often given after **surgery**, **radiotherapy** and **chemotherapy** for breast cancer to reduce the chance of the cancer coming back.

Hormones exist naturally in the body. They help to control how cells grow and what they do in the body. Hormones, particularly oestrogen, can encourage some breast cancer cells to grow.

Hormonal therapies work by lowering the level of oestrogen in the body, or by preventing oestrogen from attaching to the cancer cells. They only work for women who have **oestrogen-receptor positive cancers**.

Hormonal therapies are given to reduce the chance of breast cancer coming back and to protect the other breast. They can work in different ways and are usually given for a number of years. You'll start hormonal therapy after you have finished **chemotherapy** (if you're having it). Hormonal therapies can also be used before surgery to shrink a large cancer to avoid the need for a **mastectomy**.

Hormonal therapies are usually well-tolerated. Sometimes side effects are more troublesome in the first few months but get better over time. If you

---

4   http://www.webmd.com/cancer/tc/leukemia-topic-overview. Accessed 31.12.12

continue to have problems, talk them over with your specialist nurse or doctor, as there are ways of reducing some of the effects.[5]

Chapter Summary

The need to balance the risks and benefits, and convey them accurately, underpins every example of science communication.

As the presenter you have a responsibility to your audience to *interpret* the data as well as *report* it.

- Don't confuse information and communication.
- Think about where to start your talk to make it meaningful to the audience.
- PowerPoint is supporting you in your talk, not vice versa.
- Use the most appropriate language, and avoid inappropriate jargon.
- Be clear about what you want the audience to do, think, feel or say.
- Stress the benefits, not just the features.
- Aim for 'The Goldilocks Option' with just the right balance between complexity and simplicity.

---

5   http://www.macmillan.org.uk/Cancerinformation/Cancertypes/Breast/Treatingbreastcancer/
    Hormonaltherapies/Hormonaltherapies.aspx. Accessed 31.12.11.

3

Preparing Your Talk

The importance of preparation – the difference between a paper and a presentation – the five elements of communication planning – preparation techniques – using Twitter to clarify your messages – emotional versus intellectual audiences – an academic model of communication – the grid system of presentation planning.

Preparation: The Key to Success

The previous chapter looked at the traps which lie in wait for anyone who aims to communicate science clearly, and gave advice on how to overcome them. It was a general discussion with, I hope, clear and relevant examples. I was encouraging you to think about the point of communication, the way you do it and ask yourself whether you could do it more effectively. When I make presentations about communicating, I usually preface the material from the previous chapter by urging the audience to 'raise your head up above the daily routine and consider the point of communication', The next three chapters are a practical guide to preparing, illustrating and delivering your talk.

Let's start with:

Preparing Your Talk

All three words are crucial if you aim to be successful.

PREPARING

There are many pre-requisites for success, whether in life, in business or in specific activities like presenting or communicating. Many great men and women put preparation and hard work at the top of the list.

If people knew how hard I had to work to gain my mastery,
it wouldn't seem wonderful at all.

Michelangelo

By failing to prepare you are preparing to fail.

Benjamin Franklin

It usually takes more than three weeks
to prepare a good impromptu speech.

Mark Twain

YOUR

Notice the title of this chapter ... it's about *your* talk. Whenever you take to the stage, stand up to address an informal group or give a media interview, you are responsible for the content and its delivery. You need to own it. The timing, staging, length and topic may be in the gift of the organisers, but the time in the spotlight is your own. Turning this thought into reality can be challenging, as all kinds of obstacles appear. They include:

- It may be your talk, but it may be someone else's data or trial design.
- In corporate life you often have to pull together a talk from a range of material produced by others. You may be short of time to pull this together coherently, and make it your own.
- You may have been given the topic by a manager, and don't feel confident to deliver it to the audience in mind.
- The audience may include experts who know more about the subject than you.
- The audience may be sceptical or even hostile.

All of these problems can be overcome by the techniques outlined in these three chapters. The starting point is to prepare a talk that only you can deliver. Put some of your personality into it, and make sure you tell a story in your own way.

TALK

Many presenters have spent huge amounts of time on the data, the trial design and interpreting the results, but invest very little time in preparing the talk which will deliver it. This is understandable, as the peer-reviewed paper, published in a prestigious journal, is the cornerstone of academic research so it seems right that this should gain the lion's share of attention to detail. However, as discussed earlier, the talk is growing quickly in importance, and is your opportunity to widen your audience beyond the narrow group of academic scientists who read the journals.

A good talk and a good academic paper are like (need analogy) … they are different products of a common thread. The thread is the research and the results. The output, however, is quite different. One designed to be read, the other to be spoken.

The paper is a document of record. It will be there for all to see in decades to come, maybe for longer. For example, Edward Jenner's original paper which led to the vaccination against smallpox, and ultimately other fatal viruses, is still available:

> *Edward Jenner: An inquiry into the causes and effects of the Variolae Vaccinae, a disease discovered in some of the western counties of England, particularly Gloucestershire, and known by the name of the cow-pox.*[1]

I was able to quote from Sir Richard Doll's 1950 *British Medical Journal* (*BMJ*) paper earlier in this book, but could find no reference to the talks he gave about it. It has a permanence which talks did not have in the past, though this is now changing with so much digital storage space available. The paper is the basis for further research, and scientists today or in the future need to know all the details necessary to be able to understand, challenge or build on your work. Within reason its length is a matter for you and the journal editor.

The talk is distillation of the key information from the paper. It should establish the hypothesis clearly, outline the methods used to test it and communicate the results. It should also explain the relevance of the results to the audience. It will be limited by the time allowed by the meeting organisers, and if you go over-length you face being cut off and embarrassed. The websites of the major congresses are already archiving the main presentations, so maybe in 200 years a researcher will be able to access your presentations in the way we can still read Jenner's work … if so, I hope your descendants will feel proud of it!

---

1   Third edition. London: printed for the author by D.N. Shury, 1801. Available at: http://www.
    sc.edu/library/spcoll/nathist/jenner2.html. Accessed 31.12.11

Where the paper and the talk are similar is when you are asked to submit a poster to a conference. This is a slightly different situation from the one discussed her. Because a poster is a summary of (usually preliminary) research, what you say will necessarily mirror what is in it. Even here, however, you have the opportunity to add to what's on the poster, in the same way that you should add to what's on the slides in a main presentation.

So you've done your research, which has reached an interesting conclusion, and are ready to start writing the talk. This chapter will introduce you to a number of techniques to help you prepare effectively. I suggested in the previous chapter that you should develop a talk by asking yourself what you want to say, and what you want the audience to do, think or say after they have heard you. Notice that this is an entirely different approach from 'I'll start with the slides and talk through them.' Remember that your objective is to tell a story which will engage, educate and inform your audience.

MESSAGE MAPPING

This technique is well known so I won't spend much time on it here. It's a great way of capturing all your thoughts on a subject, then ordering them into a flow. I find it both simple and powerful. There are mind-mapping software packages available, but I prefer the simplicity of using a pen and paper. It's great for planning speeches, talks, and even books. The book you are holding in your hands now began life as a mind map.

To draw a mind map, turn the page through 90 degrees, so it is now in landscape format, and put the main subject in the centre of page. Then you surround it with the main topics, for example, safety, tolerability, efficacy, mechanism of action, properties of an ideal treatment for condition X, limitations of current treatments, key results from previous trials and so on.

I then draw arrows, or write numbers, around the sub-topics, which produces a linear flow, or a rough order. Then I take each sub-topic separately, and start a new page with that in the centre, and sub-sub- topics around it, and so on. Once I have drawn the mind map, and produced a liner flow, I turn my attention to the slides I will need to illustrate the points. There is more on the effective production and use of slides in later chapters. At this stage, I hope you can see that this technique is almost the opposite of the approach of so many presenters, who start with the slides then add the commentary to them.

MALES

This is an acronym which illustrates the five elements you need to consider:

M	Message
A	Audience
L	Language
E	Examples
S	Summary

MESSAGE

The problem with many scientific presentations is that they don't have a clear message. They have lots of data, information and so on. But as we know, information is not communication. As I was writing this I received an email from the European Respiratory Society (ERS), asking for abstracts for presentation at the annual conference. The focus of the email is not on information, but communication:

> *An unrivalled opportunity to communicate: submit your abstract today! Communicate your research, communicate your clinical experience, communicate with your peers.*

There's no doubt what the ERS thinks is important … communication! Clear communication starts with a clear message, or a small number of messages. When I lead seminars on communicating, I write this on a flipchart:

$$9 \times 1 = 0$$
$$3 \times 3 = 1$$

This usually produces puzzled looks, and occasionally a comment such as 'I know that many journalists are humanities graduates who can't do maths, but this is ridiculous!' When the laughter subsides, I explain what it means. It's an eye-catching way of expressing a simple but crucial thought: If you have too many messages, the audience will be confused. Or to paraphrase my faux equation: If you have nine messages and say them all once, the audience don't remember any. If you have three messages and communicate them three times, the audience will remember one. In reality, they may remember all three. Some years ago I was the executive producer of a TV documentary series about xenotransplantation. At the time there was exciting research being conducted in the possibility of genetically modifying pigs to enable their organs to be transplanted into humans without being rejected by the human immune system.

The series, called *Organ Farm*, was broadcast in many countries around the world, and generated considerable publicity. I was interviewed by a number of TV and radio programmes about it. Following my own advice, I developed three key messages, which I still remember, over ten years later:

- There's a major shortage of organs available for transplant. Every seven minutes, somebody dies because they can't receive one.
- You could solve that situation if you could use genetically modified pig organs, and grow them on demand in a series of organ farms.
- However, if you do that, you may be introducing dangerous pig viruses into the human race, which could cause a new epidemic against which we have no defence.

These messages were very clear, and a summary of a hugely complicated research programme. Professor David White, the scientist who led the research in the UK, and whose team bred a herd of transgenic pigs, is another great communicator. He has spent most of his working life trying to solve the problems of transplanted organs being rejected by human immune system. He summed up the challenge like this:

> There is a component of our immune system called complement. The question is, 'How does this complement identify only the foreign enemy and not destroy our own cells?' The answer is that every cell in our bodies has on its surface a set of proteins which act like flags saying to the complement, 'I am a human – don't shoot!' So these flags protect our bodies from destruction by friendly fire from complement.
>
> Unfortunately if you transplant a pig organ into a human the pig cells have flags which say 'I'm a pig' so human complement promptly destroys the transplant. What we have done is made a tiny genetic modification to the pig, which has in essence changed the flags. So instead of saying 'I'm pig' the flag says 'I'm human. Don't shoot.' False flagging has been used by war ships for centuries but it has still taken us many years to adapt the idea for pig-to-human transplants.[2]

---

2   Interview with author, clarified 14.07.11.

To reach that level of simplicity of explanation of such a complex subject requires a great clarity of thought and an understanding of the needs of the audience. It is also a great demonstration of the advice in the previous chapter, 'Make things as simple as possible, but no simpler.' Note that when I refer to something I have previously mentioned in this book, I used the phrase 'previous chapter' rather than 'last chapter' which could have you turning to the end of the book pointlessly. This is another example of 'Great communication means saying something in a way which cannot be misunderstood.'

MESSAGE CLARIFICATION TECHNIQUES

There are a number of techniques which can help you to clarify your message. Most people find they prefer one or two over the others, so I will list the main ideas here.

Technique 1: assertion – evidence – support

This is a great way to start your messaging process, and is well-suited to medical and scientific subjects. You make an assertion, back it up with the key evidence then bring in some extra supporting information. The three elements together produce a very robust message, which can then be expanded into a story flow as required.

Assertion

Before I turn to some specific examples, I would like to ensure that we all understand the same thing by the word 'assertion'.

An assertion is defined in dictionaries in a number of ways. They include 'a positive statement, usually made without an attempt at furnishing evidence', and 'confidently stated to be so but without proof; alleged'. These definitions offer a good starting point for this exercise: It's a clear, positive statement which would require evidence to prove or support it. In argumentation and rhetorical studies, it's what is called a claim (not to be confused with a 'claim' used in a legal or regulatory sense by a pharmaceutical company talking about one of its products.) Here are some examples of assertions:

- London/Tokyo is the most expensive city in the world.
- In 2010 and 2011, Lionel Messi was the world's greatest soccer player.
- A recession is when the gross domestic product of a country falls for three quarters in a row.

- Smoking causes lung cancer.
- Lowering low density lipoprotein (LDL) cholesterol reduces your risk of a heart attack or stroke.
- Cross-resistance is a major problem with HIV/AIDS medication.

The work of great scientists often results in an assertion. Newton's laws of motion are good examples:

1.   An object at rest will remain at rest unless acted on by an unbalanced force. An object in motion continues in motion with the same speed and in the same direction unless acted upon by an unbalanced force.
2.   Acceleration is produced when a force acts on a mass. The greater the mass (of the object being accelerated) the greater the amount of force needed (to accelerate the object).
3.   For every action there is an equal and opposite re-action.

One of the most controversial topics of the 1990s and early 2000s began with an assertion. It came from Dr Andrew Wakefield, a British gastroenterologist who conducted research on children who had received a triple vaccination against measles, mumps and rubella that's known as the MMR jab. His assertion that the triple jab could overload the immune system and cause the development of Chron's Disease and autism. He claimed he had evidence that children's behaviour changed drastically shortly after they received the MMR jab. 'This is a genuinely new syndrome and urgent further research is needed to determine whether MMR may give rise to this complication in a small number of people,' he told a news conference.

The MMR case is a good example to discuss, because it demonstrates that an assertion alone can be very powerful. It is designed to be an accurate, memorable summary of the main point of your argument. It is designed to attract attention, provoke discussion, maybe to promote a particular view. However, to have any scientific validity, it must be backed up by the other two components of this technique, the evidence and support. Dr Wakefield's assertion that the combined MMR vaccination could cause autism and Chron's, was not backed up by sufficiently robust evidence, and consequently lacked support in the medical and scientific community. (This is not the place to discuss its value in other communities which place a higher value on emotion and coincidence than on scientific rigour.)

An assertion alone, unsupported by evidence, is like an advertisement. There's nothing wrong with that, as far as it goes. However, if you want to persuade your fellow medics and scientists of the value of your research, you are unlikely to do it with an advertisement. You need data to back it up. Here are some assertions taken from clinical trials published in peer-reviewed journals in 2011.

- Reduction of LDL cholesterol with simvastatin 20 mg plus ezetimibe 10 mg daily safely reduced the incidence of major atherosclerotic events in a wide range of patients with advanced chronic kidney disease.
- Some of the widely practicable adjuvant drug treatments that were being tested in the 1980s, which substantially reduced 5-year recurrence rates (but had somewhat less effect on 5-year mortality rates), also substantially reduce 15-year mortality rates.
- Screening for latent infection can be implemented cost-effectively at a level of incidence that identifies most immigrants with latent tuberculosis, thereby preventing substantial numbers of future cases of active tuberculosis.
- Results of a new Cochrane Systematic Review reveal the drugs [statins] that are widely used to lower cholesterol may be of no benefit for those with no history of cardiovascular disease and may even cause more harm than good for some patients.

Evidence

The evidence that you supply to back up your assertion needs to be credible for the audience. In the medical and scientific community, this usually involves robust research which has been published in a peer-reviewed journal. Documenting the evidence in a paper usually follows the accepted Introduction, Methods, Results and Discussion (IMRAD) method. Your talk will include the key parts (but only the key parts) of the evidence.

Support

This is the final piece of the jigsaw, and ideally provides independent verification of the value of the evidence. In the case of a pivotal trial for a new drug, it may involve:

- Supporting comments from an eminent scientist, academic or physician.
- Inclusion in guidelines published by an acknowledged expert body, such as the American Heart Association, or the World Health Association, or the British Thoracic Society
- Approval from a regulatory authority demonstrating unmet need.

Technique 2: elevator speeches

The idea behind an elevator speech, or elevator pitch, is this: Imagine that you have been invited to present your research at a satellite meeting at a major congress or some other important event. After the presentation, you step into the elevator and just as the doors close, another person jumps in. This happens to be a VIP, in fact the most important person in the world, professionally, to you. They may be the chair of the world congress of your therapy area, a leading professor at an institution where you dream of working, or the head of the grant-giving institution which can fund your pet project. This person turns to you and says, 'Hi. I'm sorry I missed your talk today. What was the summary?'

The VIP is only travelling up one floor in the elevator. This means you have about 20 seconds to summarise the key points of your presentation. It's a great discipline to practise doing this. Your objective is not to explain the whole data set, but to persuade the VIP to press the 'stop' button and say, 'I'd like to hear more about that.'

Technique 3: point – evidence – point

This technique offers a number of benefits which are integral to clear communication: it puts the message up front, and involves repetition.

It works like this:

- Make your point.
- Give evidence to support it.
- Repeat your point.

As an example, assume you are about to present results of an early study of an oral treatment for Multiple Sclerosis (MS). Your introduction might talk about the need for oral treatments rather than intravenous (IV) formulations. Having established the need for an oral treatment, you will then go on to present the results. Using the Point – Evidence – Point (PEP) technique you might say this:

> **(Point)** *A diagnosis of MS as we know can be devastating for the person concerned In addition, they then find that the treatments, beta interferons, delivered by IV or infusion, involve a trip to hospital so are*

inconvenient, can be unpleasant and can have troublesome side effects. So there is a clear need for safe, effective, oral therapies which patients could just take at home.

(Evidence) *A recent study suggested that up to 44 per cent of patients stop taking the beta interferons, even though these drugs can be really effective.*

(Point) *So it's clear, then, that there is a need to develop oral treatments for MS which would be more likely to be accepted by patients. Now I'd like to turn to Drug X, one potential treatment which looks promising.*

Technique 4: Tweet my story

Earlier I talked about an elevator speech. Tweet my story is an even more rigorous stripping down of your argumentation to its bare essentials. The task is to summarise your story in a maximum of 140 characters, including spaces, as permitted by the social networking site Twitter. I am not suggesting for a moment that this technique would ever catch on in medical and scientific circles, but doing it successfully does impose a degree of clarity on your thinking. I have run this as an exercise many times, with groups including scientific researchers, with great success. The story about the need for new MS treatments would be summarised:

Up to 44% MS pts stop meds cos sfx + othr prbs. Need new oral meds. X lks prmising. Rlpses down 60%, compliance high 90% in p2 stdy.

There you have it, the key point summarised in 132 characters.

Audience

If scientific presenters could make just one change, it should be this: Put yourself in their shoes. When I present this, I illustrate it with a slide showing many photographs of many different types of shoes: high heels, pumps, workmen's boots, trainers, shiny city shoes, canvas shoes, flip flops, soccer boots and others. I also use a photo of Prince Charming trying to fit the wrong size shoe onto Cinderella. The point is well made: There is no 'one style and size fits all' for scientific presentations.

I regularly see big name, well-respected scientists deliver their standard talk on their specialist topic without any thought for the specific needs of different audiences. This is wrong, and in my view disrespectful to the audience. I recently sat in a presentation about the biology on protease inhibition where the presenter lost the audience in the detail within two minutes of a 20 minute talk which would have been more accurately aimed at post-doctoral researchers rather than HIV physicians.

This is section explores in more detail what 'their shoes' might look or feel like.

As with any population, you can divide an audience in many ways: Gender, age, language, educational attainment, wealth, geographic location, profession and religion are common ones. Before you plan your talk, ask yourself these questions about the audience:

- What does the topic mean to them?
- How much do they know about this topic?
- What preconceptions do they have about it?
- What are the barriers to them understanding your point of view?
- How much do they want to know?
- What do you want them to do after they have heard you?
- What do you need to tell them to get them to do that?

Once you have answered these questions you'll have a good impression of the audience. Then look at your data and divide it into different groups:

- essential
- nice to know
- could be included if we had more time

For example if you are talking to a conference of infectious disease experts you don't need to tell them that pneumococcal disease includes sepsis, otitis media and pneumonia, or that it is the biggest killer of children under five years of age in the developed world. However, you probably do need to update them on the changing resistance profile of different strains of pneumococcus, and where those strains have appeared in different parts of the world. The other essential information might be that details of the 13-valent vaccine being trialled, and which serotypes included in it.

Emotional and Intellectual Approaches

Apart from their scientific knowledge, there are other considerations. In particular, I would like you to consider the emotional v intellectual attributes of audiences, and also of a presentation. Some people I meet believe there is no place for emotion in a scientific presentation. However, I was in the audience at the American Society of Haematology (ASH) when the Principal Investigator received a standing ovation after he presented the results of his trial. The data set was very robust, the trial was well-designed and conducted, and the results were stunningly clear.

What feelings were evoked? Pleasure at the elegance of the trial design and clarity of results. Relief that an important medical question (could a novel drug have the same impact on overall survival in myeloma patients as an autologous stem cell transplant?) had been answered. Excitement on behalf of their patients, that there was now evidence that their lives could be extended. Yet deeper than that, was an empathy with the patients, who the physicians understand have to go through the trauma of treatment as well as suffering the painful symptoms of the disease.

The professor also exuded another emotion: passion. This is one of his defining personal characteristics, but here, presenting the results of a landmark study, his passion was even more evident.

The point about the intellectual versus emotional criteria of the audience is a partial, simplified modern version of Aristotle's ideas from 3,500 years ago. I will discuss this in more detail in Chapter 5. He believed that to communicate successfully, you need to appeal to both the intellectual and emotional sides of your audience. There has been much discussion in recent years of intelligence quotient (IQ) versus educational quotient (EQ). There is no doubt that everybody who is invited to make a scientific presentation has a high IQ. The great presenters also have a high EQ.

So how do you ensure you make both an intellectual and emotional connection? Consider a continuum from emotional to intellectual. For a moment, imagine that every audience is at one end or the other. Figure 3.1 illustrates how you would appeal to the opposite ends of the scale.

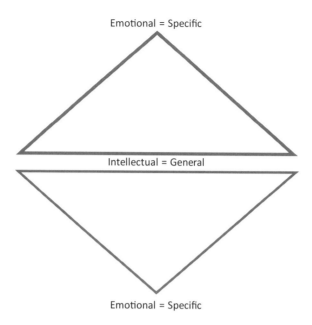

Figure 3.1   Appealing to emotional and intellectual audiences

> *To appeal to the emotions, start with a specific example then widen out to the general picture.*
>
> *To appeal to the intellectual, start with the general picture, then narrow it down to a specific example.*

Looking at this in terms of medical presentations, you might say that the bases of the triangles (intellectual) represent Evidence Based Medicine, while the apexes (emotional) are composed of anecdote and case study. In my experience, the case studies often provoke as much discussion at some medical congresses as the data.

In reality, of course, very few audiences are completely emotional, or completely intellectual (even juries in serious criminal trials have been known to be influenced by passionate pleas from skilled orators, in the face of strong evidence to the contrary). Most audiences are composed of a mixture of knowledge, experience and feelings. In other words, they are somewhere on the emotional–intellectual continuum. An additional challenge is that an audience is composed of individuals, each of whom sits somewhere on the same continuum. The challenge for you as the presenter is to gauge accurately where they sit.

An example of where this went wrong was at the opening session of the Climate Change Conference in Copenhagen in December 2009. The conference opened with an apocalyptic video, showing a young girl going to sleep peacefully but waking up to find herself in a desert wasteland. As she sets out to explore, the land on which she was standing appears to crack open and she runs away. She then faced a tornado and a flood, at which point she leapt into a tree and screamed.

The video was produced to provoke an emotional reaction, and ended with the caption, 'Please help the world'. In fact, the main reaction to the video was criticism. Many scientists and scientific commentators claimed it played on emotion, was inaccurate anyway, and ignored the factual basis of climate change. Coming as it did in the wake of the climate change email revelations from the UK, some said it was the kind of unsubstantiated rear-mongering so often quoted by climate change deniers. For our purposes, the incident offers a salutary lesson: Be sure you get the emotional/intellectual balance of your audience correct. Otherwise you harm your credibility.

Anticipating Objections

Another key part of preparing your talk is to anticipate what obstacles you will face in getting the audience to agree with you. The challenge is a classic one in communication studies. It was first exemplified by two researchers called Shannon and Weaver in 1942. They later amended it, separately and together but I now use a modified version of their original model. It looks like this:

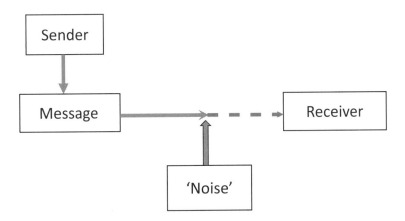

Figure 3.2   **An academic model of communication**

This is known as the Sender – Message – Receiver model of communication. It was originally devised to illustrate the challenges of using radio waves for communication in the 1940s, but its use has been widened since to include all kinds of communication particularly spoken. As you can see, it has five elements:

SENDER

If you're the presenter, you are the front line sender. However, you are probably representing many other people. Colleagues, patients, sponsors, your institution, your speciality and the conference organisers are the obvious ones. There may be others in specific cases. All of these people and organisations combine to form a general 'sender'.

MESSAGE

The previous section discussed this at length, so it should be clear.

RECEIVER

As with the sender, the receiver concept can also incorporate many and varied elements. The front line receiver is the audience, whether in the hall or online. Once again, they will represent a number of other people, groups and institutions. The receiver is an agglomeration of all of these groups.

CHANNEL

At its simplest, the channel is the communications vehicle. In some circumstances it may be a TV or radio programme, a magazine advertisement, a webcast, a website, a book or even a stage play. I made a documentary about HIV/AIDS in Africa some years ago where a voluntary group were using glove puppets to illustrate their sex education lectures. In this case, the puppets were the channel. The idea of the channel appears in many classic definitions of communication. Another famous one is Lasswell's: 'Who says what, to whom, in what channel with what effect?'

In the case of a presentation, the channel is face to face verbal communication (often with an extra audience watching via the internet).

NOISE

This is the communication element which is ignored by many scientific presenters, but is crucial to getting your point across. In Shannon and Weaver's original model, it meant anything which interfered with the perfect reception of the radio signal. It manifested itself as the interference you sometimes hear on an analogue radio. In our terms, the concept is the same. Here, it manifests itself as anything which prevents the audience accepting your point. It might be scepticism, indifference, misunderstanding, a valid point or any number of other factors. It might be, 'I don't trust this data because it's been tainted by industry sponsorship.' It could be 'I don't trust this person because I've had reservations about his/her work before.' There are many possibilities, but the effect is the same: It interferes with your signal, hence the narrower arrow in my version of the model, to illustrate that not all the message reaches the receiver intact.

Given the changes in society, and in communication techniques and technology, during the past 60 years, it would be surprising if the model in its original form had survived without criticism. The main weaknesses of it are that it assumes communication is one-way, and that the audience are passive recipients of a message in the 'Moses and the tablets of stone' sense. It is also now recognised that this is just one model of communication. There are others which illustrate new technologies such as social networking, viral marketing and similar. For our purposes, however, the model I have discussed here illustrates the main point.

ATTITUDE SOFTENING

Identifying the noise is a key part of preparing your talk. The next task is to decide what to do about it. You have two options: Ignore it or acknowledge it. The key question is whether acknowledging the noise will make it even louder, and make the audience aware of a potential objection which they would otherwise not have considered. Occasionally, this will be the case, though in the interests of transparency and within the limits of time you should acknowledge anything which is material to the case. If you decide the potential objection is worth a mention, technique from influencing skills can be very useful.

In the previous chapter I outlined a technique called 'attitude softening'. Here are some common examples of 'noise', and potential attitude softeners:

NOISE

'This study sounds fine in theory. In practice, all my patients are well treated on the existing drugs … I just don't see the kind of patients they are describing.'

ATTITUDE SOFTENER

'I understand that many physicians may feel that most of their patients are well-treated on existing medications. That may be true, but the data suggests that up to 30 per cent of patients find those medications lose at least some efficacy over time. On top of that, compliance on the existing treatments is a challenge. That's why I believe there is a real need for another treatment.'

NOISE

'There are big limitations to this study. The loading dose given in the control arm was lower than that recommended in the label. It's not a surprise the new compound did better.'

ATTITUDE SOFTENER

'One point I'd like to highlight in the study design was the loading dose given in the control arm. In the label, it recommends 600mg. However, research suggests that about two-thirds of physicians now use a loading dose of 300mg. So in the trial we let the treating physician choose whether to use 600 or 300mg. We believe this reflects clinical practice, and makes the results more relevant to the clinic.'

NOISE

'The efficacy results are impressive, but the number of patients who dropped out due to adverse events, particularly neuropathy, was huge. I think this new drug is only on the borderline of tolerable.'

ATTITUDE SOFTENER

'If we look at the discontinuations, the number of patients dropping out of the study due to neuropathy appears to be slightly high at first glance. We investigated this further, and found that 95 per cent of those patients who dropped out due to neuropathy had received a previous treatment and had suffered quite serious neuropathy, grade 3 or 4, which had subsequently resolved. That meant that they know how painful neuropathy could become on the older drugs, so as soon as they felt the first signs, they dropped out ... you can see here that most of them dropped out with neuropathy grade 1 or 2. In reality, there was hardly any grade 3 neuropathy on the new drug, and no grade 4. Grades 1 and 2 are manageable, and as far as we know, if those patients who dropped out had stayed on the drug, their neuropathy could also have been managed. Of course we can't say that definitely, but we believe that's likely.'

Language

The English poet Samuel Taylor Coleridge memorably described poetry as 'the best words in the best order'. The same could be said about a presentation. However impressive your results or clever your hypothesis, your audience won't understand it if you use the wrong words. Medical students in the UK and US learn a new language of about 6,000 words while at medical school. When they qualify and talk to patients, they have to find other words which convey the same meaning in a way understandable to non-medics. Telling a patient they have a vitamin D3 receptor polymorphism is unlikely to mean much, but telling them that they have an increased risk of skin cancer as a result of that condition may be useful information. If you are presenting to non-scientists, you face the same challenge.

If you're unsure about where to pitch your talk, take a look at some patient group websites, or a site such as www.webMD.com for guidance. However, you can do much more with language than just convey information. This applies to scientific presentations as much as anything else. Hans Rosling is Professor of International Health at The Karolinska Institute in Stockholm, Sweden. He became a minor celebrity and even had his own TV series, *The Joy of Stats* in the UK based on his energetic way of presenting statistics. You can see his original talk on the TED (Technology, Entertainment, Design) website.

What Professor Rosling illustrates is that there is a place for enthusiasm, passion and anecdote in the driest of scientific presentations. Using the right language is an important part of that. Here are some tips to add impact to your presentations:

Avoid Inappropriate Jargon

The key to using jargon successfully is to use it appropriately. If the speaker and listener understand the same jargon, then using it is no problem, and is probably expected. However, beware of inappropriate use. I once saw a professor of public health set off a minor health scare by describing two events as 'temporally related', meaning they were related by time. In this particular case, he was trying to explain that one event (a vaccination) happened before another (a baby becoming sick). What he meant to say was, 'There is no link between these two events except that one happened before the other.' Here, using the jargon phrase 'temporally related' was confusing to journalists who were covering his speech.

Another eminent professor summarised the same issue beautifully on another occasion when he said, 'You might as well run a headline that says "Man gets murdered after buying lottery ticket" … there is no connection between the events except that one happened after the other.' That, in my view, is great communication.

USE PERSONAL WORDS

Personal words are 'I', 'You' and 'We'. These are the words which connect the presenter and the audience. They invite the audience into the presenter's circle. When I make a presentation I am constantly looking for the audience to nod their heads as an acknowledgement that they recognise my point. Personal words are a way of doing that.

BE SPECIFIC

Your scientific training encourages you to be precise. In a presentation, you should follow the same rules. If you are presenting data, part of your role is to interpret the results. So you may say that the data prove, disprove, demonstrate, suggest, trend towards, support, negate or are insufficiently powered to do XYZ. Using the right word will make your presentation as clear as possible. Remember, your aim is to say things in a way which cannot be misunderstood.

Examples

The right example can illustrate a point brilliantly, or translate an abstract concept into concrete reality. It is not unusual for an audience to remember the example more than the big picture, so choose yours carefully. Examples can include typical patient case studies, details (anonymised) of your own patients, great statistics or anecdotes. The last here can be very powerful, but need to be used carefully to avoid the critical 'n = 1' attitude of some scientists.

Summary

The concept of primacy and recency tells us that the parts of a presentation most likely to be remembered are the opening and the ending. Of the two, the ending trumps the opening every time. That's why you need to make the best use of it.

The summary is your chance to do a number of things, including: Pull together a number of strands of your talk; stress the key points; introduce the consequences of your research; say what should happen next, or repeat your key messages. It should be short and focused. It should not include lots of data. It should, however, include some element of 'What happens next?' Ideally it should tell the audience what you want them to do. I will discuss this in more detail in the next section.

Turning Messages Into a Talk

So how will you put all this advice together and turn it into a talk? To discuss this, let's return to the concept of three key messages. If you develop them properly, each message will summarise or introduce a theme. Here are some common themes about new drugs which can do that:

Safety, Efficacy, Cost-effectiveness

Mechanism of Action, Drug/Drug Interactions, Compliance

New data, Differentiation over existing drugs, Clear patient profiles

Taking the first example (safety, efficacy, cost-effectiveness), you may produce these three messages:

> *Message 1:*   This new drug has fewer safety issues than the old one, in fact its safety profile is similar to placebo.

> *Message 2:*   The new drug is very efficacious … it reduced relapses by 27 per cent over two years.

> *Message 3:*   Research shows that patients on the new drug had on average two fewer hospital admissions a year, so it really does save you money.

Let's assume you are planning a presentation about HIV123, a potentially valuable new medication for the treatment of HIV and AIDS. This is a complicated field, with lots of drug combinations available, patients treated differently at different stages of their disease, resistance still an issue and many different opinions about how to tackle it. However, what is agreed by all is that the goals of treatment are:

- reducing the viral load (the amount of virus in the body);
- increasing the CD4+ count (a measure of the strength of the immune system);
- producing treatments which the patients can tolerate, and;
- avoiding resistance where possible.

In this case, your three messages might be:

1.   Efficacy: HIV123 is an effective treatment for HIV.
2.   Mechanism of Action: HIV123 has a unique mode of action.
3.   Adverse Events: HIV123 is well-tolerated.

We now have the themes and the messages you want to discuss. Now we need to build these into a talk. I recommend what we call the grid system of planning your talk.

Intro	Objective	Menu 1, 2, 3
1 :	2 :	3 :
Summary	Action	Outro

Figure 3.3   Preparing your talk using the grid system

The Grid System

PREPARING YOUR TALK USING THE GRID SYSTEM

The grid was developed by my colleague Lloyd Bracey and is a simple but very effective way of planning a talk quickly. It can be used for almost any length of presentation, and any type, formal or informal, whether you will be using slides or other props, or just having a casual chat in a coffee bar at a congress. Here is how to fill it in:

INTRO AND OUTRO

Start by filling in these two sections. The intro is, obviously, your name and affiliation, though it's better if it includes something more memorable. I'll discuss different types of impactful openings in the chapter on delivering your talk, but for now let me give you one example. I was the moderator of an HIV conference, and opened it by telling a story which I hope was relevant to the meeting, helped establish my own credentials and also made the audience pause for thought.

I said, 'About 25 years ago when I was a TV journalist in London, I was covering the emergence and rise of HIV/AIDS. At the time a professor said to me, "If you get diagnosed as HIV positive, there's very little we can do to save you. You might as well start choosing the wood for your coffin." Now, 25 years later, if two men in their 30s are diagnosed, one with HIV, the other with diabetes, the one with HIV has every chance of living longer than the man with diabetes.

That's how far we have come. That progress has been made by collaboration between industry, physicians, patients, governments, philanthropic funders like the Gates Foundation, and many others. It's worth thinking for a moment about that progress, and how successful the collaboration has been. The point of this meeting is to increase the collaboration.' Over the course of the meeting a number of people commented on my opening remarks. It's worth spending some time on trying to develop something which produces a similar reaction.

The next section to fill in is the outro. This is a word from TV news. There it's used as the final comment, as the reporter on the scene hands back to the studio. In presentations, we use it for comments such as, 'We have a few minutes for questions now, before I hand you back to the chair.' Write these in first.

OBJECTIVE

Spell out a clear objective here. In the case of HIV123, you might say, 'I want to tell you today about a new drug, HIV123, which is showing great promise in terms of efficacy and tolerability, and I hope will soon be part of our armoury for Treatment Experienced patients who've failed on earlier therapies.'

MENU

Here you introduce the key themes of your talk, which you will expand on later. Ideally you use this as an opportunity to mention your three key messages for the first time. 'I want to divide my talk into three: It's effective, so I will show you the results of the pivotal trial…. it has a different mode of action which means that resistance is unlikely to develop, and I want to show you that it's well-tolerated.'

You now have the start of your talk.

MIDDLE SECTION

The middle section of the grid forms the bulk of your presentation. This is where you present the data, and introduce other supporting material. You see that each of the three middle sections contains a number (1, 2 or 3) and three bullet points alongside it. The number refers to the theme for that section, and the bullets are supporting evidence. You need to introduce whatever types of supporting evidence will best support your case. For example, your three bullet points might be:

Bullet 1: data
Bullet 2: guidelines
Bullet 3: a case study

On other occasions your entire presentation might be data-focused, so your bullet points will pick out key points from a pivotal trial. This is the case on our presentation on HIV123. You might develop your plan like this:

Point 1: HIV123 is an effective ARV treatment for HIV

Bullet 1: It increases CD4+ count to above 200, the level at which we define as patient having full-blown AIDS.
Bullet 2: It reduces viral load down to undetectable levels.
Bullet 3: Its efficacy is sustained long term, at least up to 96 weeks.

Point 2: HIV123 has a unique mode of action

Bullet 1: It works differently from other ARVs, which means that resistance is rare so far.
Bullet 2: It targets the virus before it enters the host CD4 cell.
Bullet 3: It stops the virus entering the cell and replicating.

Point 3: HIV123 is well-tolerated

Bullet 1: In trials, the rate of AEs on HIV123 was comparable with AEs on drug XYZ, the standard of care.
Bullet 2: The incidence of renal abnormalities was lower on HIV123 than on XYZ.
Bullet 3: A study in naïve patients showed that HIV123 had a favourable lipid profile compared with XYZ.

SUMMARY

This is a summary of the key points, and in reality is another opportunity for you to repeat your key messages. If you think of a communication technique I outlined earlier of 'Tell them what you're going to say … Say it … Tell them what you said', then this is the final part of that. So here you would say 'So I hope this has been a helpful introduction to this promising new treatment, HIV 123. I hope you find the efficacy data is compelling, can see how the mechanism of action produces less resistance than earlier drugs, and find the Adverse Event profile encouraging.'

ACTION

This is where you tell them what you want them to do. In this case, you might say, 'I hope you will find it a useful addition to treatment options for Treatment Experienced patients who have failed on earlier types of medication.'

If I was giving this talk, my completed grid for this talk would now look like this:

I'm JC. Med journo. Tell anecdote re diabetes and HIV. Shows collaboration, which is point of meeting.	To introduce HIV123. Shows promise. Could become part of armoury for Treatment Experienced patients.	Three sections. Efficacy (pivotal trial) Mechanism of Action (= less resistance) Safety (AE profile)
Point 1: HIV123 is an effective ARV treatment for HIV It increases CD4+ count to above 200, the level at which we define as patient as having full-blown AIDS. It reduces viral load down to undetectable levels. Its efficacy is sustained long term, at least up to 96 weeks.	**Point 2: HIV 123 has a unique mode of action** It works differently from other ARVs, so cross-resistance is rare so far. It targets the virus before it enters the host CD4 cell It stops the virus entering the cell and replicating	**Point 3: HIV123 is well-tolerated** In trials, the rate of AEs on HIV123 was comparable with AEs on drug XYZ, the standard of care. The incidence of renal abnormalities was lower on HIV123 than on XYZ. A study in naïve patients showed that HIV123 had a favourable lipid profile.
I hope you found the efficacy compelling, can see how MoA = less resistance, and find AE profile encouraging.	I also hope you'll find it a useful addition, and consider using it for TE patients.	We have a few minutes for questions before I hand you back to the chair.

Figure 3.4   Example of a completed planning grid

The Second Grid

I also use a second grid for planning. This one is particularly useful when I am planning a story flow, or preparing for a Q&A session. Once again, it is based on the concept of three key themes. Let's stick with the example above, and see how it works. Here is the grid:

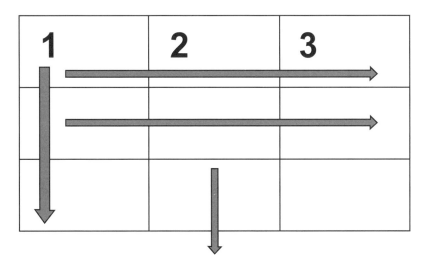

Figure 3.5   Presentation planning grid no. 2

You can see here how this grid is really a more detailed version of the middle section of grid one. In this case, though, rather than looking for three bullet points to support each main points, you go into more detail as you go down the grid. With HIV123, a completed grid might look like this:

HIV 123: A new treatment option for Treatment Experienced people with HIV/AIDS

Column 1: An effective ARV	Column 2: Unique MoA	Column 3: Well-tolerated
Increases CD4 count to >200, level below which a person has full blown AIDS.	It works differently from other ARVs, which means that resistance is rare.	In trials, the rate of AEs was comparable to those on drug XYZ, the standard of care.
It reduces viral load down to undetectable levels	It targets the virus before it enters the host CD4 cell.	The incidence of renal abnormalities was lower in HIV123 than in XYZ, the standard of care.
The efficacy is sustained long-term, for at least up to 96 weeks.	It stops the virus entering the cell and replicating.	A study in naïve patients showed HIV123 had less negative effect on lipids than XYZ typically does.

Figure 3.6   Example of a completed planning grid no. 2

USING THE GRID

The key thing to understand here is that once you have filled in the detail, you can use the grid horizontally or vertically. So you might start with the horizontal approach: 'The three things we found in the pivotal trial on HIV123 are that it's an effective ARV treatment for HIV, it has a unique mechanism of action and it's well-tolerated. As you know, these are the three key attributes of an HIV medication.'

'Now when I say it's efficacious, in the pivotal study it increased the CD4 + count to above 200. As you know, that's the level below which a patient is regarded as having AIDS, rather than being HIV positive. The different mechanism of action also showed a great benefit in the trial, because we saw no resistance at all, so you can combine it with other types of HIV medication. Tolerability is the other key point, and here we also have good news. The rate of Adverse Events was comparable to drug XYZ, which as you know is the standard of care for Treatment Experienced patients.'

You would carry on going across the columns in this way, revealing more of each columns with every step.

Or you could take the vertical approach. Here, you would take one point, for example, the efficacy results, and concentrate on that. The point is that both of these approaches work, but they suit different audiences. The horizontal approach works best with generalists, and the vertical one suits specialists. Let's move away from HIV123 and consider another example. Assume you are talking about an oral anti-coagulant to be used to prevent thrombosis after orthopaedic surgery. You are planning for three meetings, with:

1.   a physician who will discharge patients after orthopaedic surgery
2.   a pharmacologist
3.   a member of the hospital reimbursement committee

Your key messages might be about efficacy, mechanism of action and cost-effectiveness. In your talk with the physician you want to encourage them to ensure the patient continues to take the medication for three months after they return home, you would talk mainly about the efficacy, and the importance of avoiding blood clots further down the line.

When you meet the pharmacologist, you would concentrate on the mechanism of action, and how acting higher up the ras cascade of Factor Xa is a more efficient way of preventing clot formation than acting on the thrombin, further down the cascade. In your third meeting, with the member of the reimbursement committee, you would focus on the cost-effectiveness of the medication.

Grid number two will enable you to prepare for all three meetings, and others.

Chapter Summary

- The road to a successful presentation starts with thorough preparation.
- The presentation and the paper require different levels of detail.
- There are a number of techniques to help you prepare. Including:
  - message mapping
  - MALES
  - Assertion, evidence, support
  - Elevator speeches
  - Point, evidence, point
  - Tweet my story
  - Consider where your audience is on the scale of 'emotional to intellectual'.
- Anticipate objections and use attitude softeners to defuse them.
- Use the grid system to prepare your talks.

4

Illustrating Your Talk

The point of visual aids – different types – presenting without any visuals – misunderstanding PowerPoint – the three reasons behind bad slides – why the paper and presentation are different – decluttering your slides – bad slides and how to fix them – good slides and why they work – other ways to illustrate your talk – using PowerPoint to plan your talk.

Visual Aids

A successful presentation has three key elements: The content, the presenter and the visual aids. In a great presentation all three work synergistically to produce a whole which is greater than the sum of its parts. The focus of this chapter is the role of the third element: They are *visual aids*. One of the seven challenges of communicating science I outlined in Chapter 2 was, 'it's not about the PowerPoint', and this is worth repeating here. Too many presenters use PowerPoint, or other presentation software, as a crutch, not an aid. They use their slides as a parking lot for ideas, a starting point for what then becomes an unfocused talk. Other presenters read their slides, often while looking at the screen, so the presenter's voice becomes no more than a sound track for the slides. This puts the slides centre stage, rather than the presenter.

This approach fails on a number of counts:

It puts the slides centre stage, rather than the presenter. If you are going to play the supporting role in this way, you may as well just send the slides. If the words coming out of your mouth are the same as those on your slides, one of those is not necessary. Don't write yourself out of the script!

PowerPoint is designed to be a medium of impression, not information. This means that your slides are not great at conveying the details of your research.

That is best done in the published paper. However, if you have clearly divergent Kaplan-Meier curves illustrating visually how people lived longer on your new drug, a multi-coloured bar chart showing the rise of antibiotic resistance, or a pie chart showing how the incidence of diabetes or schizophrenia varies among people of different races, PowerPoint is a great way of conveying that big picture. If people want the fine detail, they can find it in the publication.

Many scientists and physicians create the problems themselves by the way they develop their talk. They start with the slides (which are often produced by someone else trying to be helpful) and work back towards a presentation. This produces the 'talking through some slides' style of presentation I criticised in the previous chapter. That is putting the cart before the horse. You need to start with the story, clarify what you want to say and only then add visual aids *where necessary* to help you say it. This is how I work with clients, whether they are preparing for a regulatory submission, a medical congress or a funding request to the board.

My colleague Lloyd Bracey is sceptical about the need for PowerPoint in every medical presentation. He tells physicians, 'If PowerPoint presentations were prescription medicines, most doctors would stop using them, because they don't work.' My view is less critical than his, but I believe there is a similarity between PowerPoint and medicines … they should both be used appropriately, in the right place and time, in the correct amount.

He also says that if you imagine that every word on a slide cost you $10, the amount of text on slides would be reduced dramatically, and be clearer as a result. See later in this chapter for examples of this.

Before I turn to the use, misuse and abuse of PowerPoint, I would like you to consider other ways of illustrating your talk. As with so many aspects of presenting, *appropriateness* is the key concept. What is appropriate for 10 people in a room may not work with 500 in a lecture theatre. Here are some ideas:

FLIPCHART

In these high-tech days, the humble flipchart seems to have lost out to its flashier cousins as a visual aid. This is unfortunate because in some circumstances a flipchart has advantages over just about any other kind of visual aid. In particular, in front of a small group it encourages interactivity and generates a sense of energy which PowerPoint often lacks. This is particularly the case if you draw on it live, in front of the group. This allows you to take the group with you as you build your story. It can be great for illustrating the flow of patients through a trial, or a timeline of events such as the key steps in a drug development programme. Use thick pens and a range of bright colours for best effect. Draw in wide, bold strokes and use all the space you have available. Although it doesn't seem a natural fit with a large audience, if you are on camera on stage you can arrange in advance to have the camera operator focus on the flipchart so that everyone can see your illustrations clearly.

VIDEO

When used appropriately (that word again!) video can be the most powerful element of your presentation. It allows you to transport the audience mentally somewhere else, or bring another person, place or thing into the hall with you. Patient testimonials and case studies often work well on video. Talking to a hall full of doctors and scientists can be intimidating anyway for people who are not used to it, so having a video recording of a patient is safer than expecting them to perform live under such stress. I once hosted a press conference at an international rheumatology conference where a German patient was one of the speakers. In the rehearsal she was fine, and spoke very good English. When we went live, however, her nerves took over, and she answered all my questions in German! Luckily her treating physician was also on the panel, so he translated.

One of the most memorable presentations I saw incorporated a dramatic video clip. The presenter was a neurologist, and he showed a Parkinson's disease patient before and after going on an experimental new treatment. The difference in the amount of control he had over his movements was hugely impressive, and the effect of the video was dramatic. It brought the point of developing new medicines into the room in a way nothing else could. One important tip if you are going to use video: Keep the clips short. Like your presentation, they should be 'as short as possible but as long as necessary'.

HANDOUTS

A bit like the flipchart, handouts are unfashionable but can work well in small groups. They are useful when you want the others in the meeting to go through detailed figures and text. Although you can do this on a slide (in a small group), giving each person their own copy on which they can make notes and comments is a real aid to understanding.

PROPS

A prop can be a great way of illustrating something. See Chapter 5 on 'delivering your talk' for more ideas and examples.

YOU!

You are the star of the show, and you can be your best visual aid. The example I quoted in Chapter 2, of cardiologist Salim Yusuf's address to more than 100 journalists at 6.30 a.m. in Chicago, is an excellent illustration. If you really want to be radical, you could try not having any other visual aids at all. I realise that this would be a brave move, but it would certainly get the audience's attention … your task would be to hold it for the duration of your talk. In reality, as I said in the previous chapter, you need to be sufficiently clear about your story anyway, and be able to tell it without visual aids in informal settings.

In reality, if you are presenting medical or scientific data to people who need to understand it you will need to illustrate it. In most cases, this will mean using PowerPoint or similar presentation software. Obviously this is not the case in chance encounters and ad hoc conversations that take place in coffee queues and around water coolers in conference centres and offices. In these circumstances you will have no help at all, which is why you need to become fluent at telling the story without any visual aids except yourself.

Before I turn to designing and using PowerPoint slides in detail, I want to clarify two points:

1. When I talk about using PowerPoint I mean presentation software of any type.
2. For the rest of this chapter I am discussing how to use slides during a presentation to a group of people. There is nothing wrong with using complicated graphs, tables and similar when you are discussing the fine details of your research with others who are closely involved. In science and medicine, the devil is in the detail and it is crucial that the detail is examined closely before you start to produce your talk.

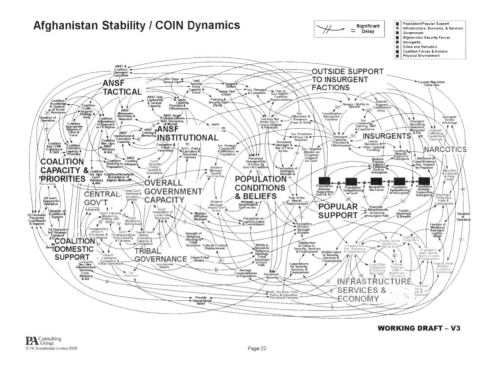

Figure 4.1   PowerPoint at its worst

PowerPoint is Innocent

PowerPoint is one of the most maligned and misunderstood tools available in the computer age. The US Army apparently banned PowerPoint in 2011 in military briefings, after the slide shown in Figure 4.1 caused confusion and derision in equal measure.

I see many complicated slides at medical congresses, and in their own way some are as confusing as this. Figure 4.2 shows some examples.

In my view, there are three main reasons for bad or unclear slides:

1. inappropriate usage
2. muddled thinking.
3. overestimating how much information people can 'take in'

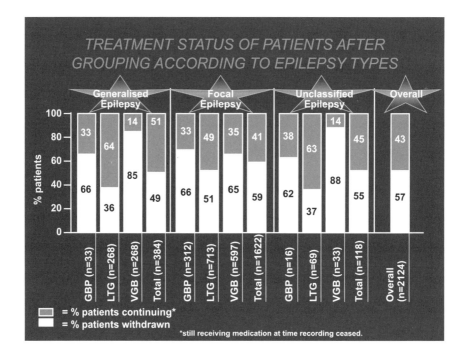

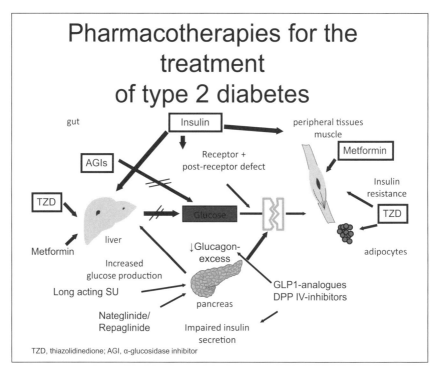

Figures 4.2   Confusing medical slides

Inappropriate Usage

PowerPoint is a very good software package, designed to illustrate presentations. Its aim is to make presentations and talks easier to understand. However, for a number of reasons, it has become the standard way of producing documents which would be better in another form, for example as Microsoft Word, Adobe Acrobat (.pdf) or Excel spreadsheets.

Many people in large companies produce PowerPoint 'decks' (to use the American term) as their basic information resource rather than as a presentation aid. They circulate these around the company for comment, and however clear they were at the beginning, the slides become busier and more cluttered. At the end of this process, someone has to use these slides as visual aids ... which by then, they most definitely are not. There are some in the audience thinking, 'Those slides look terrible!' Often, these people have had a hand in the very process!

See my comments in Chapter 2 for the disastrous illustration of how cluttered slides led to the Shuttle disaster.

Muddled Thinking

Very often, unclear slides are an illustration and a consequence of unclear thinking. People produce slides with too much information because they have not gone through the mental exercise of asking themselves, 'What does the slide need to say here to support my point or argument?' As I suggested in the previous chapter, you need to plan your talk like a story, with a clear narrative. Only then do you add in the slides. Another way to look at it is this:

> *A presentation is like a jigsaw puzzle where every piece is a slide.*
> *They all have a role to play, and must fit together perfectly.*

How Much Can People Take In?

Of necessity, data sets include a lot of information. However, that does not mean you need to put it all on a slide. It doesn't even mean you need to present it all. A basic but common error which presenters make is this: They confuse what is appropriate or mandatory to include in a clinical paper with what can be understood when it is presented on a slide. Here is my tip:

Don't confuse the paper with the presentation.

When people read a paper, they can pore over it, examine the detail and go back to an earlier section to help them understand it. In a presentation, none of this is possible. The audience has to understand it at the first pass, and as the presenter you have a responsibility to facilitate that. It is not enough to paste in graphics and charts from the paper. At the very least, they should be redrawn so they appear more visually clear.

One problem for data presenters is this:

In a report of a clinical trial, the devil is in the detail.

So how much detail should you include? Do you really need to show us the trial design, patient characteristics at baseline, their comorbidities and concomitant medications? The answer is, 'Yes, if this information is needed to enable your audience to make an accurate assessment of the study.' A data presentation is not like many other types of talk. It needs detail, and it needs to be thorough. The challenge is in deciding exactly how much detail to include. Too much and it becomes confusing. Too little and you may be accused of not being sufficiently rigorous or of concealing important information.

No two trials are the same, and there is no general advice I can give you which will allow you to get the balance right every time. Generally speaking, the patient characteristics are only worth highlighting if there is an important (not the same as statistically significant) difference between them. If you have more obese patients, or more with diabetes or compromised kidney function in one group than another, you would of course mention it. However, important differences may not be immediately obvious. I was working with a company which has produced a new type of anti-platelet agent which, according to their major trial, was significantly superior to the existing medication. About the time that the results were presented at a major European conference, claims emerged that another type of drug known as a PPI, or a Proton Pump Inhibitor, could interfere with the efficacy of anti-platelet agents. Concomitant PPI use in patients in both arms of the study then became an important factor, and was rightly included in the presentations.

On another occasion, the size of the loading dose in the comparator arm in an earlier trial was criticised. This was addressed in the later phase III trial, where physicians were left to decide which dose to use. When the phase III results were presented, this point was included in the trial design slide, and was highlighted briefly by presenters.

Think of it this way: If a company is listed on the stock exchanges, anything which could be regarded as 'material' to the stock price has to be disclosed. I urge you to take the same attitude to the fine details in your presentation … if they could become 'material', include them on a slide, and highlight them verbally.

Declutter Your Slides

Having decided how much detail is important, now look at your slides another way: how much can you fairly exclude? Cluttered slides look messy, suggest a lack of intellectual rigour and are confusing to the audience. Declutter your slides and your presentation will soar to new heights of clarity. Strip out anything which is not absolutely essential. Use the jigsaw puzzle analogy I introduced earlier to every element of every slide. Ask yourself whether you really need it on that slide, or whether it should only feature in the paper.

Here are some guidelines for producing decluttered, clear slides:

- Ensure that every slide has a clear message.
- Every slide should have a clear title.
- Title should summarise the content where possible. For example, 'Statins reduce cardiovascular (CV) events' is better than 'Relationship between statin use and CV events.'
- Every element on the slide should be necessary and clear.
- The whole deck should have a logical flow.

- Keep text to a minimum and make it large enough to read from the back.
- Use clearly differentiated colours to differentiate key elements.
- Use a light background and dark text or vice versa.
- Backgrounds and templates should be uncluttered, with only minimal logos and other identifiers.
- The design of the slides should be consistent, for example, the same colour for the active compound, comparator and placebo throughout the whole set.
- Lines and curves should be thick enough to stand out.
- Lines and curves should be clearly differentiated by colour.
- Charts (especially black and white charts) should not be pasted in from papers and journals.
- Scales should be clear.
- Spell out unfamiliar abbreviations.
- Keep references to a minimum.
- Use phrases, not sentences.
- Add arrows or boxes to highlight key points, for example, difference on Kaplan-Meier curves.
- Avoid excessive numbers of fonts and colours.
- Use sans serif fonts, for example, Verdana, Arial, Helvetica and Tahoma (as approved by the FDA).
- If you have to use a complicated slide, use animation to build it up and take the audience through the logical process. Watch TV news to see how they build up quite complicated graphics quickly, then follow their lead.

Unclear Slides

I now want to turn to typical errors which people make when they produce slides for medical and scientific data. This section contains some slides I have seen presented at meetings and congresses, and others which are available on the internet. In some cases I have changed figures and other key details to spare the blushes of those involved. In all cases my comments are aimed at the design of the slides and how much information is on them, rather than the details of the information itself. I do not intend to list every PowerPoint data sin here, but to give a few examples to illustrate common faults and suggest improvements.

Background and Prior Work

Current statin guidelines emphasise the need to achieve specific goals for LDLC to maximise clinical outcomes.

However, accumulating data indicates that statin therapy has greatest efficacy in the presence of inflammation and that statins reduce the inflammatory biomarker hsCRP in a manner largely independent of LDLC.

Further, in both the PROVE IT – TIMI 22 and A to Z trials of patients with acute coronary ischemia treated with statin therapy, the best clinical outcomes occurred among those who not only achieved LDLC < 70 mg/dL, but who also achieved hsCRP levels < 2 mg/L. In both of these trials, even greater clinical benefits accrued when hsCRP levels were further reduced below 1 mg/L.

Figure 4.3   Background and prior work

Background and Prior Work

These prior data are consistent with the understanding that atherothrombosis is a disorder of both hyperlipidemia and inflammation, and that statins have anti-inflammatory as well as lipid-lowering properties.

Despite the consistency of these data, whether achieving lower levels of hsCRP after initiation of statin therapy is associated with improved clinical outcomes in a similar manner to that associated with achieving lower levels of LDLC remains highly controversial.

We prospectively tested this hypothesis in the large-scale XYZ10 trial.

Figure 4.4   Background and prior work

The Problem

These are two introductory slides which explain the background to a trial examining whether reducing high sensitivity C Reactive Protein, a marker of inflammation, in addition to reducing low density lipoprotein cholesterol (LDL-C), would reduce cardiovascular (CV) events more than reducing LDL-C alone.

The two slides commit some of the most common faults:

- This information presumably came from the publication abstract: The author has confused the paper and the publication.
- There is far too much text, making the slides cluttered and difficult to read for the audience.
- The text contains full sentences. This makes it almost certain that the presenter will read the slides out, reducing their role to an audio track to the slides, rather than remaining as the dominant figure in the presentation.
- This amount of text will take too long to read out. The audience can read silently faster than the presenter can read aloud (if they want to be understandable), and will therefore have reached the end a long time ahead of them. Many presenters realise this, so speed up to make it less obvious. This detracts from their authority.

How to Improve Them

Reduce the amount of text.

Use phrases not sentences. For example:

Do statins work best when inflammatory markers present?

Trials suggest greatest outcome benefit with biggest hsCRP reduction:

- *PROVE IT – TIMI 22*
- *A to Z*

Target reduction: <1 mg/l hsCRP

This approach leaves the speaker to elaborate on what the slide shows, and sets the scene neatly and briefly.

Conclusions

- In treatment-experienced patients, including those with NNRTI resistant virus, ABC125 was superior to placebo
    - 59% of patients achieved confirmed undetectable viral load (<50 copies/mL) with ABC125 + BR at Week 24

- Even in the absence of any other fully active background agents (PSS = 0), with ABC125, 45% of patients achieved undetectable (<50 copies/mL) viral load
    - response rates increased as more active agents were used in the BR

- Better responses were achieved in patients with higher CD4 cell counts and lower viral loads for both treatment arms
    - higher responses were apparent with ABC125 compared with placebo, for all categories of baseline viral load or CD4 cell count

- 13 ABC125 RAMs were identified
    - an increasing number of ABC125 RAMs was associated with a decreasing response in both treatment arms
    - in the ABC125 group, the greatest added benefit was seen with <3 ABC125 RAMs
    - 86% of patients had <3 ABC125 RAMs

- ABC125 demonstrated significant activity and provides a new treatment option for patients with resistance to other NNRTIs

BR = background regimen; RAM = resistance-associated mutation;

Figure 4.5   Conclusions

The Problem

This is the conclusion slide from a presentation about a new HIV treatment, which I have called ABC125. It was presented to HIV specialists, so the high number of acronyms, common in that therapy area, is not a problem. It is interesting that even then the authors found it necessary to explain two acronyms which they thought may not have been understood (Background Regimen (BR) and Resistance Associated Mutations (RAM)).

The problem here is that there is just too much text on the slide, and it is too dense.

How to Improve It

Split the conclusions across three slides, headed:

1. Conclusions: Efficacy
2. Conclusions: Resistance
3. Summary

The last slide should contain one clear message about what it means to physicians and patients. If appropriate it may also contain a reference to further studies.

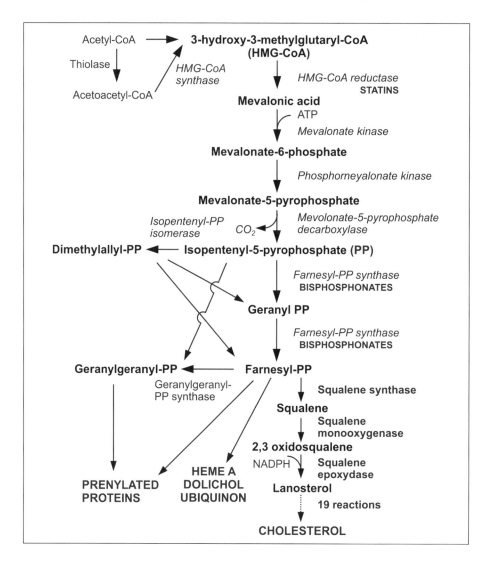

Figure 4.6   Hyperlipidemia slide

The Problem

This was part of a complicated presentation on hyperlipidemia I saw presented to cardiologists. It is too complicated. The slide, and much of the content, would have been better suited to a biology lecture where the lecturer could talk the students through all the steps in the process. As it is, I looked around the hall and saw that not many of the audience understood it.

How to Improve It

Given the level of detail and the mismatch with the audience, I would suggest not using it all in this setting. If it has to be included, it needs simplifying as much a possible. Even then I would suggest using animation, revealing the steps one at a time to aid understanding.

Remember the earlier tip: Put yourself in their shoes!

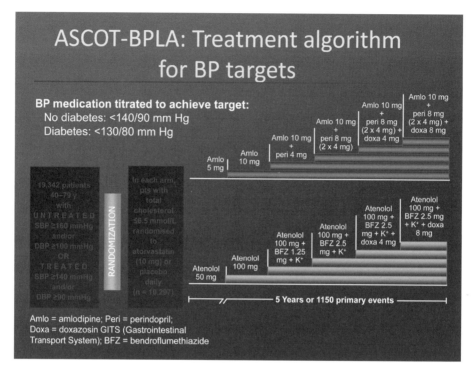

Figure 4.7   ASCOT

The Problem

This was part of an introductory presentation about the Blood Pressure Lowering arm of ASCOT, a landmark study in the prevention of CV events. It is hard to read because the colours are too dark and too close: the original has a dark purple background and dark blue text. The steps graphic on the right does a decent job of illustrating in impressionistic manner how the treatments were stepped up as patients needed more medication to keep their blood pressure (BP) under control. However, the panels on the left make it too cluttered and hard, not to say impossible, to read. As it stands, the drug and dosage details are too small to read, and the timeline along the bottom is confusing: were patients titrated to achieve the BP targets, as the caption on the left claims, or at specific time intervals, as the timeline suggests?

Confusion like this takes the audience's minds off your talk, as they try to work out what is right.

How to Improve It

The details on the panels on the left could have been on a separate slide, leaving more room for the steps. This would then allow the text to be legible. The drug could have been differentiated by using different colours for the abbreviations. The conflict between the timeline and the claim about 'titrating to target' needs resolving.

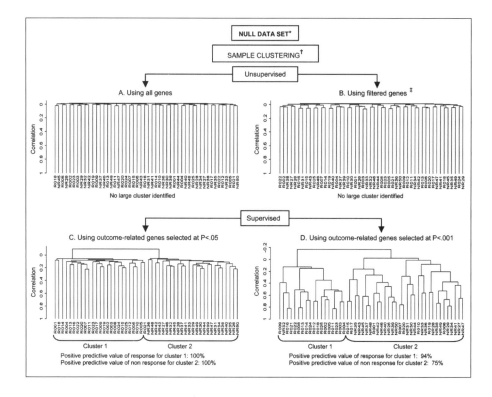

Figure 4.8   Sample clustering

The Problem

This is a classic example of a graph or table from the paper being pasted in to a PowerPoint slide and becoming illegible. It contains far too much information and contains no clues as to what is important.

How to Improve It

Don't paste tables or graphs in from the paper unless you really need to. It looks terrible, is illegible and adds nothing to the audience's understanding. Presenters do it because it saves time, but in my view this is counter-productive. However painful it is, get it redrawn in a more visual way. Before you do that, ask yourself if you really need to include all of this information in the presentation. If there is nothing remarkable about any of it, strip it right back. You need to prioritise your information for the audience. If you really have to include all this information you need to find a way of highlighting what is important. Circle the key parameters, use arrows or highlight the relevant lines to make them stand out. Put alternate lines in a different colour so our eye can move across easily.

	Newer Regimen		Standard Regimen	
	Amophine	5mg	Acenooil	50mg
Step 2	Amophine	10mg	Acenooil	100mg
Step 3	Amophine	10mg	Acenooil	100mg
	Ferindoprill	4mg	BFZ-K	125mg
Step 4	Amophine	10mg	Acenooil	100mg
	Ferindoprill	8mg	BFZ-K	25mg
	Amophine	10mg	Acenooil	100mg
	Ferindoprill	8mg	BFZ-K	24mg
	Doxazosin GTS	4mg	Doxazosin GTS	4mg
Step 6	Amophine	10mg	Acenooil	100mg
	Ferindoprill	8mg	BFZ-K	25mg
	Doxazosin GTS	8mg	Doxazosin GTS	8mg

Figure 4.9   Newer v standards regimens

The Problem

This list of drugs and doses is illegible even at close range. What is that column on the left, highlighting different steps? Where is the title and references? What is meant by 'newer' ' and 'standard' regimens?

How to Improve It

This is a classic of its kind. It's a complex chart that should never have made the leap from page to slide. If anybody wants to pore over this amount of detail they can do it by reading the paper, not during your presentation. If you feel you have to include it for completeness, ask yourself the questions: 'What is this slide telling us? What is the point of this piece in the jigsaw?' You are going to present it, redraw it in appropriate colours, or shade alternate lines so the eye can follow it. If there is anything unusual about the content, highlight it.

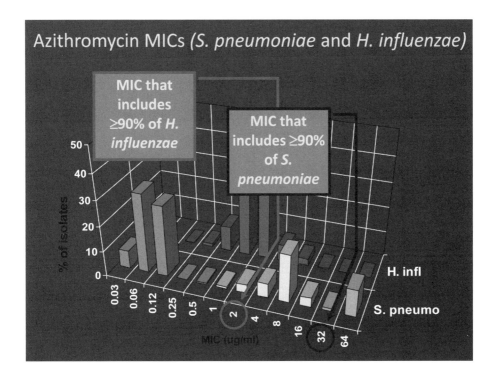

Figure 4.10   Azithromycin

The Problem

This is trying to demonstrate the Minimum Inhibitory Concentration (MIC) (that is, the least that works) required of an antibiotic called azithromycin on two common pathogens. There may be good information here, but the slide has been taken over by the Pointless PowerPoint Procedure, that is, the 3D graphics. In addition, the right hand label is covering the top of one of the 3D towers and anyway, I'm not sure what this slide is trying to say.

How to Improve It

Leave the fancy graphics out, and revert to a 'normal' format of bar charts or graphs. Have a clear message.

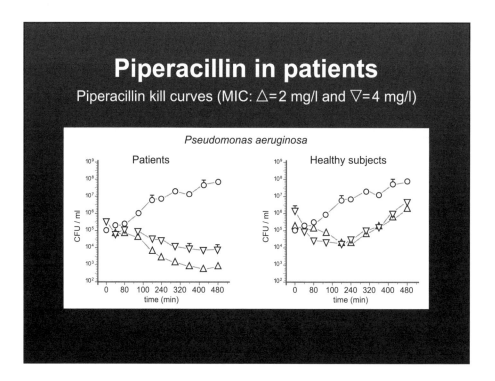

Figure 4.11   Piperacillin in patients

The Problem

CFUs are colony-forming units, a measure of viable bacterial (or fungal) numbers which are an indication of the efficacy of an antibiotic at preventing them replicating into large colonies. CFU/ml is an accepted way of counting them. The difficulty with this slide is that, because the table has been pasted in from a paper, it is indistinct. The lines are too thin, the distinction between the inverted and right way up pyramids is not great enough, and there is no clue what the circles are.

How to Improve It

Redraw it in bold colours, add a proper legend so we know what's what. Make the table a bigger part of the slide.

Good Slides

Here are some good examples of medical data slides. They all contain very clear combinations of text and graphics.

ISIS-2: Second International Study of Infarct Survival

Purpose
To assess the effects of intravenous streptokinase (SK) and oral aspirin, alone and in combination, shortly after suspected acute MI

Reference
The ISIS-2 collaborative group. Randomised trial of intravenous streptokinase, oral aspirin, both, or neither among 17,187 cases of suspected acute myocardial infarction: ISIS-2. *Lancet* 1988; ii: 349-60.

Figure 4.12i   ISIS-2: Second International Study of Infarct Survival

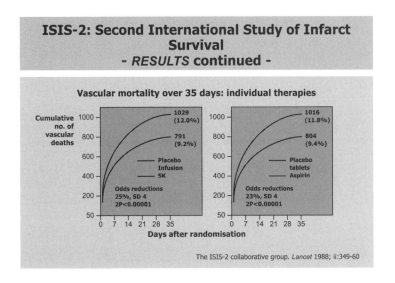

Figure 4.12ii   ISIS-2: Second International Study of Infarct Survival – Results continued

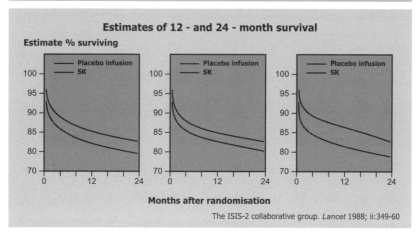

Figure 4.12iii   ISIS-2: Second International Study of Infarct Survival –
Results continued

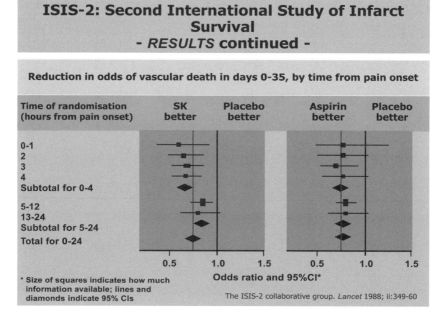

Figure 4.12iv   ISIS-2: Second International Study of Infarct Survival –
Results continued

ISIS-2: Second International Study of Infarct Survival
- *RESULTS* continued -

Effects of SK and aspirin on clinical events

Clinical event	SK allocation		Aspirin allocation		Combination therapy	
	Placebo infusion (n=8595)	SK (n=8592)	Placebo tablets (n=8600)	Aspirin (n=8587)	Both placebos (n=4300)	SK and aspirin (n=4292)
Reinfarction	202	238	284	156	123	77
Major bleed (transfused)	18	46	33	31	11	24
Minor bleed (not transfused)	81	297	163	215	33	167
Stroke (excluding TIA)	67	61	81	47	45	25

The ISIS-2 collaborative group. *Lancet* 1988; ii:349-60

Figure 4.12v   ISIS-2: Second International Study of Infarct Survival –
Results continued

ISIS-2: Second International Study of Infarct Survival
- *SUMMARY* -

Early therapy with SK and aspirin in patients with MI:

- Individually and in combination significantly reduced all-cause mortality
- In combination demonstrated addictive effect

Figure 4.12vi ISIS-2: Second International Study of Infarct Survival –
Summary

Why They Work

These are some of the results from ISIS-2, a landmark study of CV drugs which was published in *The Lancet* in 1988. The results were dramatic, as you can see from the clear difference in the Kaplan-Maier curves. What I like about this entire slide set is the overwhelming clarity over every aspect, from the title to the summary. It is a textbook example of how to translate complicated trial data into compelling, easy-to-understand slides. The colours, text and graphics all work together to leave nobody in any doubt about the significance of the findings.

The way the authors have placed the aspirin, streptokinase and placebo groups side by side, and used the differences to produce a compelling narrative, is exceptional. Notice also the clarity of language on the summary slide. A lesson to us all!

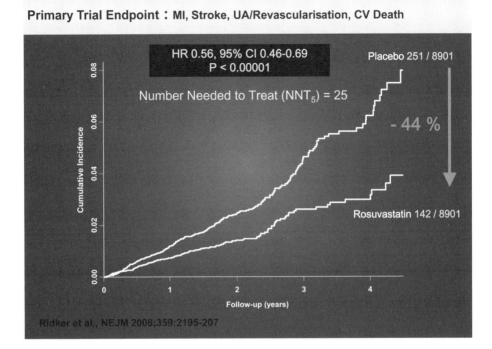

Figure 4.13   Primary trial endpoint

Why it Works

The primary endpoint of another major trial into the reduction of CV events. Everything you want to know is on here, and I like the way the implication of the odds ratio of 0.56 has been spelt out: A 44 per cent reduction in myocardial infarction (MI), stroke, unstable angina, revascularisation, and CV death. The arrow pointing out the difference between the Kaplan-Meier curves is a great asset to understanding.

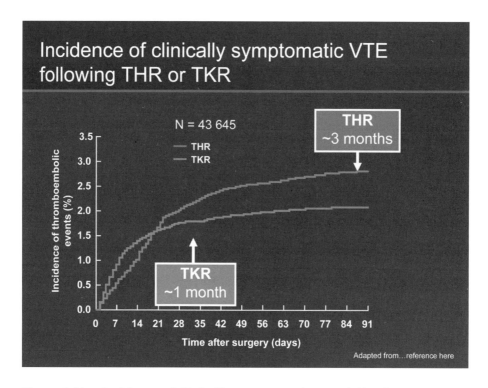

Figure 4.14   Incidence of clinically symptomatic VTE following THR or
TKR

Why it Works

This is a graph showing the different risk periods for Venous Thrombo Embolisms (VTEs) after Total Hip versus Total Knee Replacements. The clear background, thick lines, labelled axes and uncluttered look make it impossible to misunderstand and give the speaker lots of opportunity to add their own comments.

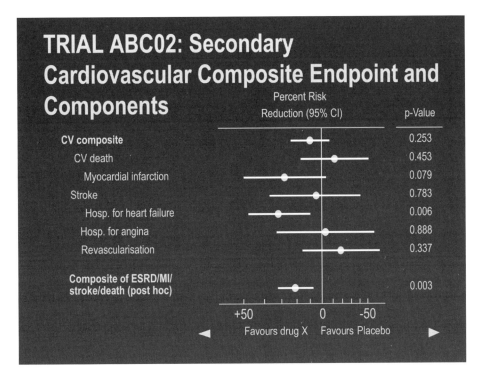

Figure 4.15   Trial ABC02: Secondary cardiovascular composite endpoint
and components

Why it Works

Forest plots are complicated and often confusing when they are on a slide.
Too often they are just pasted into the presentation from the paper. This
excellent example shows why it is good to redraw them in bold shapes, strokes
and colours. Everything on the slide is important – there is no extraneous
information.

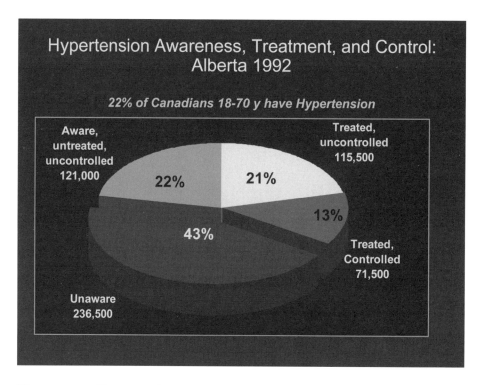

Figure 4.16   Hypertension awareness, treatment, and control: Alberta 1992

Why it Works

A simple pie chart in bold colours demonstrates clearly that more than 4 out of 10, almost half the adults aged under 70 in Alberta, Canada with hypertension don't know it, and only 13 per cent have it under control. A technique like this is so much more effective than a list of figures.

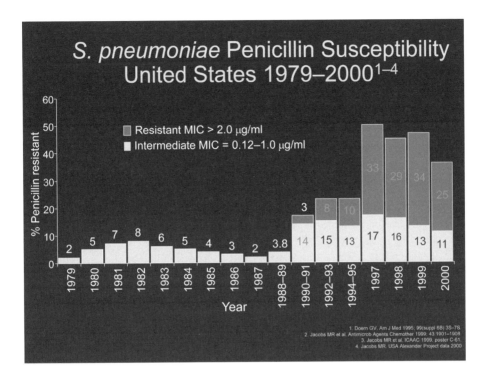

Figure 4.17   S. Pneumoniae Penicillin Susceptibility United States 1979–
              2000

Why it Works

Another slide about the problem of antibiotic resistance. This one shows the
alarming rise in penicillin-resistant streptococcus pneumonia in the US. I like
the two-colour shading to show the strength of resistance.

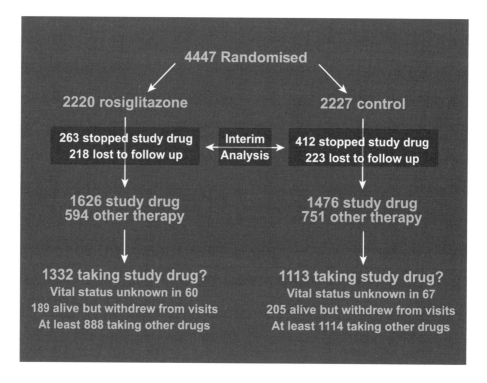

Figure 4.18   Randomised trial design

This slide was presented at an FDA hearing into alleged safety concerns of a diabetes drug. The trial design here is very easy to follow, as are the questions raised by the speaker.

PowerPoint as a Planning Aid

I want to end this chapter by introducing you to one way in which PowerPoint can be a real advantage: as a tool for helping to prepare your talk. In Chapter 3 I outlined a number of techniques including mind-mapping. I then suggested that you turn your mind map into a linear story flow. PowerPoint can help you do this. Here's how (The precise instructions may differ slightly depending on which version you are using. The sequence below is correct for PowerPoint 2010):

1.   Open a PowerPoint presentation.
2.   On the 'view' menu click 'normal'.

3. On the left hand side you will see two tabs: 'slides' and 'outline'. Click 'outline'.

This shows you the text from all the slides in your presentation. The title of each slide is in bold type. Imagine the titles are a storyboard for your presentation. If you use them properly, anybody should be able to understand the overall flow of your talk by reading the slide titles alone. To ensure this you need to follow another of my tips: The title should summarise the content of the slide, rather than set up a hypothesis. As one example, a title:

Statins reduce the risk of cardiovascular events

Is a better title than:

Relationship between statin use and rate of cardiovascular events.

To use the storyboarding technique, start building your presentation by filling all the slide titles. Once you have finished you can then go back and fill in the details on each slide. Here are the slide titles from a presentation about Type 2 Diabetes and managing hyperglycaemia in developing countries:

Title: Managing Type 2 Diabetes in Developing Countries

- High prevalence of T2D in developing countries
- Socio-economic factors drive T2D prevalence
- Diabetes can damage eyes, kidneys and nerves
- T2D has a major impact on CVD
- CVD mortality is high in people with T2D
- Hyperglycaemia undiagnosed in CAD patients
- T2D common in acute MI patients
- Hyperglycaemia induces vascular damage
- Early detection of hyperglycaemia reduces CV risk
- Controlling post prandial hyperglycaemia is crucial
- Screening: A useful tool
- Screening criteria: International guidelines
- Therapeutic options for T2D
- Lifestyle and pharmacological options
- Oral diabetes therapy: Benefits
- Oral diabetes therapy: Risks

- Oral diabetes therapy has good safety profile
- Summary

You can see from this how it is possible to follow the overall flow just by reading the titles. I hope it has illustrated just how simple and powerful the storyboarding technique can be.

Chapter Summary

- PowerPoint can be a real benefit if you use it properly. The slides are your visual aids, not vice versa.

  There are three main reasons for unclear slides:

  1. inappropriate usage
  2. muddled thinking
  3. overestimating how much detail people can absorb

- The clinical paper and the presentation require different levels of detail.
- Just because you know it all does not mean you have to say it all.
- Follow the tips in this chapter to declutter your slides. In particular:
  - Avoid full sentences.
  - Do not paste in tables and other graphics from clinical papers.
  - Use colours, arrows, circles and other graphic devices to aid clarity.
  - Split text across two or more slides if necessary … or reduce the amount.
- Avoid fancy PowerPoint tricks that obscure the facts.
- Use the PowerPoint slide titles to produce a storyboard and to help plan your talk.

5

The Performance: Delivering Your Talk

Why you should present data – the trio of presenting – vocal, visual and verbal elements of a presentation – Aristotle's influence on twenty-first-century presenters – powerful openings – the importance of signposts – adding impact with props and quotes – interacting with your slides.

The Point of Presenting

The title of this chapter addresses a key element of making presentations that many medics and scientists ignore: It's a performance. In the final planning process before I am going to speak at a conference or congress, my clients sometimes ask, 'Can you send me your presentation?' I reply: 'No, but I could send you the slides.'

The 'presentation' is made up of me and my visual aids (which often includes slides, but also flipcharts, video and props). As I pointed out in Chapter 2, 'It's not about the PowerPoint'. When I say this to scientific audiences, some of them say (and I guess that more of them think but don't say), 'Yes, but you don't understand. My audience are only there for the science. They don't want all these fancy presentational tricks. It's all about the data.' This is a fair point, but in my view it is misguided. I have taught presentation skills to thousands of scientists and medics all over the world, and almost without exception they have agreed that presenting the data in a lively, interesting fashion is better than doing so without any thought for the style and the audience.

If all the audience want is the data, let them read it in the peer-reviewed journals. Why should they bother coming to your presentation? Why do 30,000 cardiologists and their associates attend the American Heart Association every year if they can get the same benefit by reading the journals? Of course, there are other benefits to attending, particularly networking opportunities, but the primary reason is to hear colleagues speaking about their research, which is the subject of this book. They also attend because they get the opportunity to ask questions about it.

If your presentation is really just your slides, ask yourself why you should bother going. Why should the audience bother going? Just send the slides and they will get what they need. Before you plan any presentation, ask yourself, 'Why am I doing this as a presentation, rather than just sending an email or letter?' The answers include:

- I can put something of myself, my own personality, into the presentation which would not be present in a published paper.
- I can explain the findings in a way which is tailored to the audience's understanding, knowledge and expectations.
- I aim to be memorable, so that the audience remembers the key points.
- I can answer questions and clarify misunderstandings.

I want to go back to the objection raised earlier this chapter, and the idea of 'presentational tricks'. This is a phrase I have heard occasionally from people who genuinely believe that I am trying to get them to be dishonest with the data. They also use phrases such as 'style not substance' and 'spinning the results'. To clarify my point here: This is not what I am suggesting. The greatest crime in science and medicine is to play fast and loose with the facts. For decades, there was a suspicion of scientists who could communicate clearly to non-scientists, as though they had broken a code of silence imposed on the scientific community, or that if they could talk about it clearly, they were not 'proper scientists'. For years the British Medical Association actively discouraged its members from appearing on TV programmes and explaining medicine to the public.

Those days are long behind us, and scientists such as Brian Cox, Robert Winston, Hans Rosling and Richard A. Muller, the American physicist and author of *Physics for Future Presidents* are media superstars as well as holding prestigious academic posts. I am not suggesting 'presentational tricks'. My point is actually a deeper one. An audience is composed of individuals, who all have lives outside the auditorium. In those lives they watch TV and movies, tune in to webcasts and podcasts, and are exposed to highly sophisticated communication techniques. As a consequence, when they turn up for your lecture they expect you to talk them through your findings like a TV presenter, explaining, clarifying and contextualising the key points as you do so. They also want you to be enthusiastic about your subject. This chapter is the final piece in the jigsaw to help you achieve these objectives.

Question: What can the ancient Greeks tell us about presenting data in the twenty-first century?

Answer: Quite a lot, actually.

The ancient Greek philosopher Aristotle is regarded as the father of rhetoric, defined by Webster's Dictionary as 'The art of speaking or writing effectively.' Rhetoric can also be considered as the art of argumentation, or 'the art and study of the use of language with persuasive effect'. You can find many variations of this online, but for our purposes here, this will suffice. Note that rhetoric, as I use it here, has nothing to do with the modern use of the word, as in 'empty rhetoric' or 'that's just rhetoric!' or indeed 'a rhetorical question'. From the time of the Ancient Greeks to the mid-nineteenth century, rhetoric was regarded as one of the 'three ancient arts of discourse', along with logic and dialectic. They were an essential part of training for anyone being preparing for public life or any role which involved public speaking. Teaching of the ancient arts of discourse is no longer popular, but the underlying techniques are as powerful as ever. They are also great tools for powerful presentations, and have particular relevance to scientists and medics.

According to Aristotle there were three means of persuasion: ethos, logos and pathos.

ETHOS

An appeal based on ethos relies on the acknowledged character and standing of the speaker or author. It means that the presenter is a renowned expert on the subject, their authority is accepted by the audience. It also implies an element of respect from the audience to the speaker. A close modern concept is the phrase, 'ethical appeal'.

As a scientist or medic who has been asked to present at a conference or congress, you already have the right ethos to appeal to the audience.

LOGOS

This gives us the modern word 'logic'. An appeal based on logos follows a logical argument. It means persuading by reasoning. In science it is the most important method of developing a case. It also suggests that the arguments and logic are clear to the audience, well-explained and supported by facts which themselves are supported or have been demonstrated.

Your clinical or scientific research, the subject of your presentation, was based entirely on logos.

PATHOS

This is the most misunderstood of the three terms. An argument based on pathos is often taken to mean one based on emotion. TV advertisements are based on pathos. In reality, the appeal of pathos runs deeper than just emotion. It suggests a feeling of 'I know what you feel.' It means identifying with the audiences viewpoint, and is closer to the modern word, 'empathy', than to raw emotion.

A key difference between your published paper and your presentation is the element of pathos. The empathy with your audience is a key element of success in your presentation.

Ethos, logos and pathos should not be regarded as separate elements in you presentation: it is better to think of them as being fused together, so that the whole is greater than the sum of the parts. The presentation from the Spanish professor I mentioned in Chapter 3 which received a standing ovation did so because the praise was based on all three elements: His trial had produced stunning results which showed that novel agents had great promise in the treatment of Multiple Myeloma (logos); his own standing and reputation in the haematology community was already very high, and was enhanced by the study results (ethos); he understood the challenges of treating Multiple Myeloma in elderly patients, which the audience dealt with every day (pathos). The fact that this eminent professor (ethos) had demonstrated (logos) how new treatments could help the practising physicians (pathos) was a perfect illustration of the relevance of Aristotle's ideas to our world.

The Trio of Presenting

So much for the theory. In practice, how do we take translate that into a powerful scientific presentation? Many of the techniques required are common to all types of business presentations, and are grouped naturally under three headings. I think of them as the trio of presenting:

1.   body language
2.   voice
3.   content

These are sometimes referred to as the 'vocal, visual and verbal' elements of a presentation. Here are the key criteria:

Body language	Voice	Content
Authority	Volume	Relevant
Energy	Clarity	Structured
Passion	Variety	BME
Gestures	Pauses	Signposts
Look at them	Pitch	Summary
Eye contact	Pace	Conclusion
Smile	Enthusiasm	Action Call
	Repetition	

Body Language: Essentials

AUTHORITY

Nobody enjoys watching a nervous presenter, and the first thing your audience want to feel is that you are comfortable in this situation. This requires feeling confident in a number of aspects: Your subject overall, the material you are going to present, and the reaction you anticipate in your audience. The previous chapters should enable you to achieve all three. The question is: How do you demonstrate authority? By appearing confident. An audience will make up their mind about you within the first few seconds, so make sure you start right. Slow down, be measured, wait until the chair has finished introducing you then walk briskly onto the stage. Shake their hand if appropriate, look up at the audience and smile … it suggest you are looking forward to what's coming, welcomes them, shows you are human and crucially, relaxes your facial muscles.

Using the other techniques outlined in this chapter will create a feeling of authority.

ENERGY

After authority comes energy. Think of the powerful presenters you know, and consider what sets them apart. One factor is the energy they put into the performance. From the moment they bound onto the stage you can sense their physical energy, which acts like a lightning conductor to the audience … it energises them all and makes them sit up in anticipation. Isn't that how you want your audience?

PASSION

This is something which is unique to verbal as opposed to written communication, and applies whether you're communicating face-to-face, via video link or the internet. It applies in a number of ways, from the passion a really great presenter feels for their subject, to the passion they show in telling you about it. Hans Rosling's presentation at TED in California in 2007 is a great example of both.

GESTURES

Two words are crucial in getting this part right: *Appropriate* and *inclusive*. On stage in front of 1,000 people your gestures can be expansive, with wide, sweeping arm movements and exaggerated facial expressions as you demonstrate confidence and ownership of the stage. In front of half a dozen colleagues in an informal review meeting you would generally be advised to tone down the histrionics. We've all been in meetings where someone who doesn't understand the form goes 'over the top' and it does their credibility no good. The converse is also true … if you don't display an animated personality on the big stage, it can swallow you up and you'll find yourself less than memorable.

An *inclusive* gesture is one which includes the audience, such as opening your arms out wide with palms facing outwards. An exclusive gesture is one which excludes all or part of the audience. Turning your back on the audience to point at the screen with a laser pointer is an exclusive gesture. It might also mean that they can't hear you, as you go 'off-mic'.

On the subject of laser pointers, it is worth noting here that they do have a place in the presentation of medical and scientific data when used appropriately. This means pointing out key parts of a table, curves or forest plots, for example. They are indispensable when showing images of tumours, X rays, angiograms and similar diagnostic tools. The secret is to use the laser where needed, and to point clearly at the target. Avoid the 'laser whirls' we see so often, where the presenter feels they need to use the laser on every slide. You don't need it on most slides.

It isn't only your own gestures that are important. When I make a presentation I am also constantly looking for one gesture from audience members: A nod. When people nod it suggests agreement, and gives me confidence that they have got my point. I then work to make them nod again to my next point, and so on.

LOOK AT THE AUDIENCE

We use video analysis in our training courses. The most common fault (and most surprising one, to the attendees) is the amount of time they spend looking at the screen, rather than at the audience. It is important to realise here that a data presentation does pose particular challenges. The main one is that in some circumstances you want to *discuss* the data, rather than just *present* it. This usually happens in a small group of colleagues, fellow investigators or other peers. It is very common in slide review meetings, where the presenters go through their proposed presentations with others involved. In these situations it is perfectly acceptable, even expected, to turn to the screen and talk, and pick on specific data points to discuss.

What is not acceptable, however, is to do that in your main presentation. Here, you need to refer to the key data, and ensure the audience know what you are trying to tell them. However, for most of it, you should face them, not the screen.

EYE CONTACT

'The eyes are windows on the soul.' 'I want to see the whites of their eyes.' 'Look me in the eye and tell me this won't go wrong.' 'He looks shifty, and refused to look me in the eye.' There are many expressions relating to the importance of eye contact. They all apply when you are making a presentation. I usually identify the people at the extremities, front row, left and right, middle row on the end, and the ends of the back row, and make sure I look at them as I make my case. If the audience is too big, or the lighting is too low so you can't see them, this is not possible. In these circumstances I make eye contact with those at the front, and where appropriate I occasionally walk to the front of the stage, shielding my eyes from the light, to make contact with people at the back.

I also try to synchronise my eye contact with the rhythm of my speech. In this way I move my head, looking at different audience members, in time with my key points. My own presentational style is quite informal and conversational, so I regularly ask them, 'Do you know what I mean?' or 'Do you find the same thing?' These are moments of high eye contact.

SMILE

Appropriate smiling says you're human. It's perfectly human to smile, and when you do it, it relaxes your facial muscles, shows you are relaxed and encourages the audience to do the same.

Voice: Essentials

CLARITY

The first vocal rule of spoken communication is that you must speak clearly. This means enunciating the words, and ensuring that the audience can hear every syllable. In particular, beware of contractions such as 'hasn't' rather than 'has not' 'doesn't' rather than 'does not' and 'didn't' rather than 'did not'. In data presentations, these are key words, and if you say them unclearly, the audience may misunderstand. The phrase 'superiority over the current treatment hasn't been demonstrated by this large randomised trial' is a key point of your presentation. If half the audience think you said, 'has' then they'll be struggling to understand what comes next. You need to say, 'superiority over the current treatment has *not* been demonstrated by this large randomised trial'.

Figures and percentages are another area where clear enunciation is key. I heard a presenter last week talk about a 'thirteen per cent increase in PFS' the first time she referred to it, then a 'thirty per cent increase in PFS' the second time. The slide she was presenting was so complicated it took me a few seconds to find the correct number. While I was doing so, and thinking, 'Is it 13 or 30?' I was not concentrating on the other points she was making. If she had said it clearly I (and I assume many other audience members) would not have been confused.

There is a simple thing you can do to ensure your presentation passes the clarity test: Record yourself practising it. Ideally, record yourself on video, use your mobile phone or other handheld device if that's all you have to hand. If you can't bear looking and hearing yourself back, you're in good company. Most people, when they first hear or see themselves, hate the way they look and sound. You'll get used to it, and in this instance it will definitely be worth it ... no pain, no gain!

VOLUME

Many years ago, when microphones, amplifiers and speaker systems were unusual, nervous presenters would often begin their speech by asking the audience, 'Can you hear me at the back?' Today, when even quite small gatherings use an amplified Audio-visual (AV) system, 'being heard at the back' is rarely a problem. However, there is more to 'volume' than simply being heard. Just as we make quick decisions about someone based on their body language, we do the same on their voice. Projecting your voice is another way you convey confidence and authority. To project effectively, your voice needs to come from your diaphragm, not your throat. Note that this is not the same as shouting! Practise projecting your voice in big halls, and know the difference between this, which is acceptable, and shouting, which is not.

PACE

Speaking too fast is a common fault of presenters, yet I have never heard a presenter speak too slowly. I hear them speak without enthusiasm, mumble or be indistinct, but never too slowly. English-speaking TV presenters speak at 180 words a minute, or three words a second. If speaking too fast is your problem, try this exercise:

Write a script of 360 words, exactly as you would say it out loud. Use contractions where appropriate, such as it's, we're and you're (and avoid them where inappropriate, as outlined above.) Record yourself reading it. It should take two minutes. In fact, if it contains a lot of long words, it may take slightly longer. It should not take less. If you finish too quickly, repeat the exercise until it takes you two minutes. It may feel really slow to you, but that is the reading speed of a TV newsreader, and is a good guide for presenters. It's like when the golf or tennis pro wants you to hit the ball differently. To do that successfully you need to know what it feels like when you get it right. Just as the golf pro wants you to feel the 'click' of an iron on the ball, or the tennis pro wants you to experience the sweet feeling of the sweet spot, I want you to feel the correct reading speed. Then you need to make it a habit.

VARIETY, PITCH AND PAUSE

These aspects of the voice are best considered together, as each is an aspect of the other.

Listen to professional broadcasters or actors and you will find that whatever kind of voice they have, they all share the attribute of introducing variety into their speech. This is particularly true of radio broadcasters and actors. (If you are in the UK, listen to *The Archers* on BBC Radio 4 for great examples.)

Mozart said, 'The rests are as important as the chords.' In the language of the lay person, he meant, 'The gaps are as important as the notes.' Think of a child learning to play the piano, and what do you hear? The gaps are in the wrong place, or too long or too short. The rhythm is wrong. A common failing with presenters is that they don't use enough pauses, so their speech patterns become monotonous. Use pauses for dramatic effect, to let your words sink in, to ensure understanding when addressing a multilingual audience, to make your speech more interesting.

PITCH

Generally speaking, someone with a high pitched voice has to work harder to establish their credibility than someone whose voice is deeper. Many politicians, from the former British Prime Minister Margaret Thatcher to the Chancellor George Osborne, have realised that a high-pitched voice can make them sound shrill and grates on the ear. If you have a really high-pitched voice, record it and play it back, however painful that may seem. Then try lowering the pitch gradually. If you are going to do a lot of public speaking, seek professional help from a voice coach.

ENTHUSIASM

Listen to a good radio interview, or a presentation from one of the scientists I mentioned earlier, such as Professor Brian Cox or Hans Rosling. They convey their enthusiasm with their voice. Marcus Du Sautoy, the Professor of the

Public Understanding of Science at Oxford University is another great example of enthusiasm. They do it by using the techniques here, of variety, change in tone and pitch, and appropriate pausing. Listen to them and analyse how they do it.

REPETITION

This sub-heading could easily be in the 'content' column but I think of it as a vocal aspect of a great presentation. The use of repetition is a major difference between a written paper and a verbal presentation. Use it for emphasis, to underline your main finding, or to ensure the audience understands it. So you might say, 'The Kaplan-Meier curves here show the difference at the 18 month time point. This (point to relevant place) represents a 44 per cent drop in the risk of a cardiovascular (CV) event or CV death on the new drug compared with placebo. A 44 per cent drop. That magnitude of difference is unprecedented and has never been demonstrated with any previous randomised placebo-controlled trial.'

You would never write it like that in a scientific paper, but saying it like that adds emphasis and variety.

Content: Essentials

RELEVANT

Like so much of this book, I am writing this section on a long haul flight. I predict that when the Captain comes on the PA system to speak to the passengers as we start our approach to Newark, New Jersey, he will say something like, 'It's a lovely day down there, with a temperature of 29 degrees and a wind speed of 15 knots from a north westerly direction means we'll be landing this afternoon on the south runway.' The wind speed and direction are crucially important to him (I'm delighted that he knows them!) but to me they are irrelevant. I want to know whether the flight will be on time, whether we have to get a bus to the terminal, and what flight number to look for in the baggage reclaim area.

I see the same in many presentations. The presenter has a 'standard spiel' which they deliver, irrespective of the audience. I have talked a lot about relevance in this book so I won't labour the point any more except to repeat what I said earlier: 'Put yourself in their shoes.'

STRUCTURED

A presentation is like a journey. It has a start and an end, and some points of interest along the way. Some points are more important than others. It helps us to know what they are, why they are significant and when to look out for them. A good structure facilitates this. Use the grid system of presentation planning outlined in Chapter 3 to make sure your presentation has a good structure.

SIGNPOSTS

Without signposts we would find it hard to know where we are on a journey. It's the same in a presentation. We use signposts to signal a new section of the presentation, or a change of direction. They help to divide up our speech into discrete chunks. They are also an invitation to people whose attention has wandered to rejoin us, and refocus on the presentation. Here are some signposts I have heard recently:

> *'So that's the efficacy data on the 20 mg dose. Now I'd like to turn to the 30mg results ...'*

> *'So that's a summary of the drug's impact in the liver. But what does it do to the kidneys? That's what I'd like to look at now. ...'*

> *'The efficacy, as we saw there, is better than the current treatments. But is this achieved at a cost of safety ...?'*

SUMMARY

I referred in a previous chapter to an old yet effective communication technique:

- Tell them what you're going to say.
- Say it.
- Tell them what you said.

The summary is where you 'tell them what you said'. It is a short summary of the key points. Depending on what you have presented, it may include the background to why you developed the drug, that is, one short point about the unmet need. However, it is not an opportunity to reopen the whole issue, or re-present the data. Ideally it should reflect our opening remarks.

CONCLUSION

The conclusion takes things one step further than the summary. It puts the findings into context and tells the audience what it means to them. So the summary might be, 'I hope my talk today has shown that there is now a new treatment option for people with X condition, and that you will consider it for appropriate patients.

ACTION CALL

In the world of advertising, the 'Call To Action' is so important it is often known only by the acronym CTA. In that world, huge amounts of money and energy are invested in deciding which words work best in a CTA. There is lots of interesting research on it, suggesting that if you want people to phone in and buy a TV-advertised DVD or piece of jewellery, you should say, 'Our advisers are standing by right now for your call', and if you want to get hotel guests to recycle their bath towels you should say, 'Sixty-eight per cent of the guests who stayed here before you decided they didn't want their towels washed every day.'

Our CTA is simpler. In fact it is so simple it is often forgotten (this would be a serious offence in advertising!). In Chapter 2 I urged you to 'start with the end in mind'. Now I offer you the other half of that advice: 'End by telling them what you want them to do.' So if your data demonstrates that there is now an effective oral treatment for Multiple Sclerosis for the first time, tell your audience that you would like them to consider prescribing it for suitable patients. If it suggests a new drug is not safe for patients with renal failure, suggest they avoid them. If your conclusion is that you now need to run a large multi-centre trial, ask them to enrol patients in it.

Powerful Openings

The British Olympic athlete Linford Christie won the 100 metre gold medal at the Barcelona Olympics. He dedicated the medal to his coach with the words, 'He got me to go from the B of the Bang.' Going from the B of the Bang of the starting gun is clearly crucial for Olympic sprinters, but it also important for presenters. Research suggests that an audience makes up its mind about someone within just 17 seconds of them starting to speak. The experiment has never been done in learned or scientific circles, and you probably have a little more leeway, but the general message still holds true: First impressions last.

So how do you ensure you make a powerful start? Here are some attention-grabbing techniques, with real examples from my own experience:

BOLD STATEMENT

Make a statement which is not only bold, but makes them think. It may be a prediction. I knew a researcher in the field of xenotransplantation who would begin his annual lecture for first year medial students with the words, 'In the early decades of the twenty-first century, when you are practising physicians, medical science will be able to enable many people to live to 120. But when they die, their bodies will be 30 per cent pig.'

The most powerful opening may be a personal statement. I was working with a pharmaceutical client on their presentation to the European regulatory authority for marketing authorisation for a new drug for multiple sclerosis (MS). The authority assessors had produced an extremely negative report about the drug's safety, efficacy and whether there was really a need for it at all. The main presenter was a distinguished Italian neurologist who had worked in the MS area for decades before he joined the company. To return to the 'ethos, logos, pathos' criteria, he had bags of ethos. He clearly felt that the drug had been unfairly assessed, and in a break from rehearsals he said to me, 'I don't recognise the picture that the assessors have painted of this drug.'

I said, 'Let's start with that, then. That's our opener!' His colleagues took a little convincing, but eventually agreed. On the day, he started his presentation like this, 'Before I turn to the data, I would like to make a personal statement. I have been in this field for more than 35 years, and have seen many potential treatments come and go. I do not recognise the picture of this drug which has been painted by your assessors' report.' It shook the regulators out of their normal PowerPoint-induced mood, and made them sit up and listen.

DRAMATIC FACT OR IMAGE

I was running some presentation skills training courses for European psychiatrists. For his final exercise, a Dutch member of the group said he wanted to practise presenting to members of the police force in Amsterdam. He began it with this stark statistic and image: 'Good evening, I am Dr ... I'm a psychiatrist and I specialise in treating people with schizophrenia. In Amsterdam, this city in which we all live and work, there are about 1,500 people with schizophrenia. Because of the nature of their illness, you as police officers may come across some of them in your work. You may find them doing strange things, causing a nuisance, being abusive or just walking down the street shouting. I want to explain to you tonight that they are not criminals, but are ill. I also want to help you to deal with them.'

This was a great opening, combining a number of different techniques:

- an inclusive statement, which joins us as a group (this city in which we all live and work)
- a dramatic fact (1,500 people with schizophrenia)
- an image (being abusive, shouting)
- a clear outline of his talk (explain they are not criminals, and how you can deal with them)

A MEMORABLE STATISTIC

I was hosting a debate on asthma in Morocco, and opened my introductory remarks with this: According to the international guidelines, almost everybody with asthma should have normal or near-normal lung function, should never or hardly ever be admitted to hospital due to an exacerbation, and can all lead a normal life. Why is it, then, that this year 250,000 people around the world will die of asthma?'

Statistics abound in data presentations … choose one that makes people sit up. Notice in the asthma example I did not say 'every year 250,000 will die …' I deliberately set the statistic in the future tense, 'this year, 250,000 people will die …'. This is more dramatic, and conveys a sense of 'why can't we prevent this?' which the audience shares.

A QUESTION

Ask a controversial or thought-provoking question to get the audience's attention. For example:

When I was a boy, we all thought robots would be making our lives easier by now, and that space travel would be the norm. Whatever happened to the robots and space ships?

When resources are as scarce as they are now, what justification is there for carrying out cardiac surgery on obese smokers, or liver transplants on alcoholics? These people have brought it on themselves. That's my topic today …

INCLUSIVE STATEMENT

A technique used by the Dutch psychiatrist mentioned above. This is a statement which brings people together, and acknowledges their common causes, challenges or activity. It may be a simple reference to the difficulty of so many people getting together with so much pressure on their time, or a mention of travel difficulties overcome. I opened a haematology conference in Amsterdam on the day after the city has suffered its worst-ever snowfall, and the airport had been closed. My opening remarks made reference to the enormous efforts that people had made to get there, and hope that they and their patients would find it worthwhile. Or it may be a reference to medical or scientific challenges, for example 'Multiple Myeloma is still incurable. But the skill of physicians like you in this room and the availability of new drugs have made it a chronic condition.'

A PROP

A prop in this sense is a physical object which demonstrates something. It may be a revolutionary inhaler, or a pill which cost more than a Jumbo Jet to develop. You will see a great example on TED talks, where British inventor Michael Pritchard demonstrates his device for making filthy water drinkable.[1]

I attended a pain management conference in London where the presenter used a prop to great effect. It was a small meeting of about 25 medical journalists. On the tables in front of each of us was a big clip which we use for holding papers together. We call them bulldog clips in the UK. The speaker introduced herself: Good evening, my name is Dr … from the University of London. My subject is pain management. On the table in front of you, you'll see a bulldog clip. Can you please pick it up, and clip it onto your finger? Thank you.'

She continued introducing herself and after about 30 seconds said, 'How does the clip feel? It hurts, doesn't it? How much does it hurt? On a scale of 1 to 10? When I ask you that question a number of thoughts probably come into your mind. You probably think, 'How much is it supposed to hurt?' Or 'How much does she think it hurts?' This illustrates the first challenge of pain management – pain measurement.'

---

1   http://www.ted.com/talks/lang/eng/michael_pritchard_invents_a_water_filter.html. Accessed 01.01.12.

You may already use props with your patients to illustrate body parts and how they work. Think about what props you could introduce into your presentations to help understanding and help them become memorable.

A QUOTE

I have used a number of quotes in this book, and I have used some of them to open presentations. The quote can be short and pithy, or longer and thoughtful. Either way, it should be attention-grabbing and bear close scrutiny. If you choose the right quote, it can make the audience think, and set the tone for the rest of your presentation. I was asked to speak about the reputation of the pharmaceutical industry to an audience of pharmaceutical executives. Many of them had heard me speak on previous occasions, and I wanted to signal right at the start of the speech that this would be new and different. I opened with a quote from George Merck, who was President of Merck & Co from 1925 to 1950. He was known as a thoughtful, compassionate man who also had a great instinct for business. Speaking at Virginia Medical College in 1950, he said this:

> *We try never to forget that medicine is for the people.*
> *It is not for the profits. The profits follow.*
> *If we have remembered that, they have never failed to appear.*
> *The more we remember it, the larger they have been.*

I opened my presentation with this quote, then posed the question, 'What would George Merck have thought of some recent pharmaceutical industry practices, which have been uncovered in the media and featured in court cases?' I listed some of them, and urged the audience to look at them from the point of view of an outsider. I acknowledged the difficulty of improving the industry's reputation but urged the executives to work hard to achieve it. It will be worth it in the end, I said. I ended my presentation with a famous quote from President John Kennedy, about why it is worth attempting difficult things:

> *We choose to go to the moon in this decade and do the other things,*
> *not because they are easy, but because they are hard, because that goal*
> *will serve to organize and measure the best of our energies and skills,*
> *because that challenge is one that we are willing to accept, one we are*
> *unwilling to postpone, and one which we intend to win*

Despite delivering an unpalatable message, my speech was very well-received.

INTERACTING WITH YOUR SLIDES

A great presentation is a coming together of the presenter, the visual aids and the audience. In this final section I want to give some tips for the way you interact with your visual aids. In the majority of cases, this will mean PowerPoint slides, so I will concentrate on using them.

USE SIGNPOSTS

The 'signposting' technique I outlined earlier in this chapter is particularly important when interacting with your slides. You should say, 'So that's the data on the drug's effect on the liver. There has been some concern expressed about potential effects on bone formation, and that is where I would like to turn now.' Then, and only then, do you reveal the next slide, on bone markers. It is important to do it in this order, because the audience will concentrate on what you are saying for those few important seconds. If you reveal the slide then say it, they will be trying to figure out what's on the slide, and may miss what you say.

PAUSE AND INTRODUCE THE SLIDE

You are very familiar with your slides, but each one comes as a surprise to your audience. Once you reveal a new slide, the first thing you must do is to pause and explain the layout. This is particularly important in the early part of the presentation, when you are familiarising them with the look as well as the content. You need to explain the axes, the scales and the colours. You might say, 'The vertical axis is the drug concentration, the horizontal one is time (in hours) since last dose. The blue curve is the control group, and the yellow one is the treatment group.'

TELL THEM WHAT'S IMPORTANT

Given the subject matter, and however much we try to simplify it, some slides still end up looking quite complicated. Building up a slide can help here, as I outlined in Chapter 4. Even then, you need to tell the audience what is important. When presenting Kaplan-Meier survival curves, you might say, 'If you look here you will see that the curves separate quite quickly, and by the end of the study there was an overall survival difference of 5.4 months. This is why the trial was stopped early. The ethics committee decided it was not ethical to keep patients on the placebo group, when those on the treatment arm were gaining such a big benefit.'

If you are presenting the design of a number of trials, you might say, 'What is important here is this figure: the percentage of patients who continued from one study to the next … 93 per cent. That's a great vote of confidence from the patients, as it's a measure of their benefit and satisfaction with the experimental therapy. One would hope, though obviously we can't prove this yet, that it may lead to improved adherence in clinical practice.'

One difficulty for an audience unfamiliar with the details is to understand whether a big or a small number is beneficial, or whether a reading should go up or down. Sometimes a trial includes some values you want to increase, and others you want to reduce. Simple examples are that you generally want to reduce blood pressure and low density lipoprotein (LDL) cholesterol, but increase high density lipoprotein (HDL) cholesterol and T-scores on a bone mineral density test. You want to normalise blood glucose and INR levels. It can be disconcerting to an audience, when all the previous slides have contained values that have gone up from baseline, when you suddenly present a slide where they all go down. There's a simple solution to this … tell them! Make it clear what you are aiming to do, and what success would look like when measured on this parameter.

USE THE COLOURS

This is a simple but powerful technique. Repeat the key colours to aid understanding.

Chapter Summary

You present data for a number of reasons, including:

- to be memorable
- to add your own personality
- to add your credibility to the data
- to tailor your explanations to the audience
- to answer questions and clarify misunderstandings

Good communicators are no longer regarded with suspicion by scientists.

Aristotle believed that there are three pillars of argumentation, which are still relevant today:

1.   ethos
2.   logos
3.   pathos

The three elements of a great performance are:

1.   body language
2.   voice
3.   content

Great opening techniques include:

- bold statement
- dramatic fact or image
- memorable statistic
- a question
- a prop
- a good quote

Use signposts throughout your talk to break it up.

When interacting with your slides:

- pause and explain it
- tell the audience what is important
- highlight anything surprising
- put the information into context
- point out colours to help understanding

6

Medicine and Science in the Media

The popularity of science stories – media challenges for scientists – different types of journalist – general versus specialist correspondents – social media's influence – the role of news agencies – the journalist's world – the importance of relevance – making it understandable – making it interesting – avoiding 'So what?'

Media Challenges

The media has an insatiable hunger for stories about science and medicine. There are more doctors and scientists interviewed on TV news than any other profession except politicians and police officers. If you want to communicate widely about science and medicine, the media is the most powerful tool. In this chapter I will sketch out the media landscape and highlight the parts of it you are likely to encounter. Every tip, technique and guideline in the book so far is relevant to media interviews. However, for the scientist, the media poses additional challenges.

The agenda of news journalists and scientists are often in conflict. Journalists can be truth seekers, but are first of all story seekers … that's what they get paid for. Much news journalism is black and white, whereas much science is grey. Where scientific advance is usually incremental, journalists seek the 'Eureka!' moment. Many journalists typically ask 'Is it A or B?' when in reality the answer may vary from A to B, sometimes be AB and on occasions be C, D or E. A challenge for scientists is that in much of today's non-specialist media, fact and opinion are combined (sometimes in the same article) to produce powerful, emotional stories which stray a long way from the basic facts of the case.

Take one example of truly ground-breaking science: the cloning of Dolly the sheep at the Roslin institute in Scotland in 1996 (but not reported until February 1997).

The news was due to appear first in *Nature*, the ultra-respectable scientific journal. The title of the paper was typical of scientific reporting and gave no clue as to the drama behind the breakthrough, or the media frenzy which was to ensue:'Viable Offspring Derived from Fetal and Adult Mammalian Cells.'[1]

The Lancet editorial described it as:

> ... *a technological achievement of great interest. This is the first report of a live birth containing genetic material from an adult mammalian somatic cell that was manipulated in what are now fairly standard embryological techniques. The Roslin group have done what was thought to be impossible; the DNA of an adult cell can no longer be looked on as having gone along an irreversible path of gene suppression during differentiation into a mature somatic cell. Roslin lamb 6LL3 ('Dolly'), now over 6 months old, was derived from DNA from a mammary epithelial cell of a 6-year-old ewe in late pregnancy.*

The editorial also stated:

> *Cloning a human being remains as far away in practice as it ever did.*[2]

The *Nature* embargo didn't quite work, as the UK newspaper *The Observer* acquired the story from another source. The newspaper's science editor broke the story on the front page on 23 February 1997, days before the planned publication in *Nature*: 'Scientists Clone Adult Sheep – Triumph of UK Raises Alarm Over Ethical Issue.'[3] Notice here how, even in a 'serious' newspaper, in a story written by the highly-respected science editor, there is an extra element of controversy introduced by the angle of 'ethical alarm'. However, that was nothing compared to the treatment the story received from some of the UK's more sensationalist newspapers:

---

1   Wilmut, I. et al. *Nature*, 385: 810–813.
2   'One Lamb, Much Fuss'. *The Lancet*, 349(9053): 661.
3   McKie, R. *The Observer*, front page, 23 February 1997.

Fears Over Sheep Clone
Scientists have 'bred' the world's first cloned adult animal – sparking
fears that a woman could give birth to her father's twins. Other horror
scenarios include a dictator replacing himself, and showbiz moguls
cloning dead stars.[4]

'Age of the Man-made Monster?'[5]

Given such stories it is easy to see why the idea of talking to journalists can make even experienced scientists worry. However, media coverage is where most people hear of scientific advances (House of Lords Select Committee on Science and Technology, 2000), so it is worth persevering. One reason for scientists' reticence is that the media itself, in some countries at least, has a terrifying reputation. Many doctors agree with the former British Prime Minister Tony Blair, who said the media can be 'a feral beast, tearing people and reputations to bits'. He was speaking from years of experience, and there is no doubt that in some cases he is absolutely right. However, here is a comforting thought for you: You will not be subjected to that kind of treatment from journalists.

Rather than looking to discredit you, journalists are more likely to ask:

- Can someone explain this to me, and help me explain it to my audience?
- What does this mean to them?
- Why should I care?
- How can I check it?
- How can I sell this to my editor?

These are the questions you must answer. Before we get into detailed discussion, I would like to clarify some of the terms I use in this chapter:

When I use the word 'audience' I mean it to include readers, viewers, listeners, internet followers and users of social media sites. In most developed countries today, journalists are what is called 'multi-media operators', and write for all the outlets which their employer organisation offers.

---

4   *The Sun*, English and Welsh editions, 24 February 1997.
5   *Daily Mail*, 24 February 1997.

When I talk about 'news media' I mean journalism aimed at the general public, for example the long-established daily newspapers, TV and radio shows and their internet relations. If I talk about the 'scientific media', I mean specialist publications such as *Nature* magazine, or the medical journals including the *Journal of the American Medical Association* (*JAMA*), the *New England Journal of Medicine* (*NEJM*), *The Lancet* or similar publications aimed at particular specialities such as the *Journal of Clinical Oncology* or *Blood*.

Specialist Correspondents

This leads to an immediate question: What about the science and health correspondents of large respected media outlets, for example, *The Times*, *New York Times*, *El Pais*, *Le Monde* and similar publications, or TV news programmes? Where should they be categorised? Generalists or specialists? Their role is complex, and can vary hugely from one organisation to another, depending on their own backgrounds. Generally speaking their role is to understand enough of your complex science to be able to report it accurately and interpret it for their audience, in language the audience understands. To find the answer, you need to look at the articles published under their name. I use this term deliberately, because the process of production, especially in newspapers, involves sub-editors whose job is to write the headlines and cut the story to fit the available space. You can see immediately that this raises two issues:

The reporter or correspondent doesn't write the headline. I will discuss headlines later in this chapter. For now, take note of one fact: In countries which have a press complaints body, there are usually more complaints about headlines than about the body of the story.

The sub-editor is a generalist working on a specialist's article. The sub-editor (known in journalism as a 'sub') may have little or no knowledge of the subject, but has to make the piece fit the available space. This may just involve cutting it from the bottom, or it may involve rewriting it. It is easy to see where mistakes creep in, even with the best intentions. In addition, the subs are sometimes asked to take two similar but unrelated stories and combine them into one piece. This is not uncommon with reports of a medical or scientific congress, for example where two set of data on the same topic are presented on the same day. If one set of data is yours, be prepared for some frustration in the way it is reported!

It is also important to understand that the term 'specialist' does not necessarily mean 'specialist' in any meaningful way that you would recognise. If you are a medical or scientific specialist, you will have studied your subject for years, published on it and be regarded as having expert knowledge and insight. I was in a meeting recently where a new member of the group introduced herself as 'being here because I'm a specialist in platelet biology'. Over coffee she told me that after her Ph.D. (on platelet biology) 'I then began to specialise in megakaryopoiesis and haemostasis. That was just over ten years ago.' She was, by any definition, a specialist.

A specialist correspondent, on the other hand, may come from a variety of backgrounds. Some, particularly science correspondents, will have at least a science degree, though nothing more (and nothing like the level of expertise of the woman I mentioned above). Others, particularly medical correspondents, may have come into the field from news journalism, having enjoyed writing about science and medicine and shown an aptitude for it. Very occasionally, the medical correspondent may be a physician. In addition, consider for a moment the professional areas of interest of a medical correspondent:

- All the illnesses, diagnostic and treatment options known to medical science, from acne and Alzheimer's to X-rays and Yellow Fever.
- New drugs and procedures (often including homeopathy and Chinese medicine).
- Medical controversies such as stem cell research, cloning and even in some cases, debates over assisted suicide. I saw a piece very recently by a medical correspondent in the UK headlined 'Dead or Alive? Not Always Easy to Tell.'
- Health politics, including the consequences of government changes, new regulations, finance, doctors and nurses pay, cutbacks and private versus public funding issues.
- Health scares, such as the debate over the safety of the Mumps, Measles and Rubella (MMR) jab in the UK, Bovine Spongiform Encephalopathy (BSE), thiomersal in medicinal products, fluoridisation of water and leukaemia clusters around electricity pylons.
- The occasional 'exotic' diseases which suddenly appear, such as the Ebola virus, anthrax poisoning or necrotising fasciitis.
- Pandemic coverage, most recently avian and swine flu.

With an in-tray like that, it isn't surprising that when this journalist meets the platelet biologist, the conversation is hardly a meeting of minds … and then, when they have written the piece, the sub-editor gets to work on it.

The best starting point when planning to deal with journalists can be summed up like this:

> *Never underestimate their intelligence.*
> *Always underestimate their understanding.*

Other Types of Journalists

If you are planning to engage with the media, you are most likely to come into contact with specialist reporters and correspondents, as outlined above. However, a wider understanding of the media world and its inhabitants will help you to make the most of the opportunities offered.

The first point to consider is this: The internet has changed journalism fundamentally. Mainstream journalists feed off social media, which in turn repackages, recycles and comments on articles in the mainstream media.

The most obvious consequence is that the traditional 'gatekeeper' model, where the media are gatekeepers allowing or preventing access from the scientific community to the wider public, is now largely redundant. There is a wealth of 'gatekeeper' research in communication studies, much of it looking at who controls the gate and why some stories pass thorough it while others fail. I have conducted some of this research myself, looking at the news production processes of an international TV news agency. If I was to do more, I would start with Twitter and Facebook, and widen out from there.

Thanks to the internet, news organisations and journalists are no longer in charge of the news. In fact, now that social media is so pervasive, the very idea of 'a journalist' has changed. Traditionally, this was a person who earned their money from researching, writing and publishing stories in a recognised publication. They were separate from members of the public (or other interested parties such as scientists) whose main contribution to the media was to pay for it. Now, anyone can be a journalist. I imagine that many of the readers of this book have their own blogs and Facebook pages. Many of you will be on Twitter. You probably belong to various discussion groups based on your own area of interest. In that case, you already have the basic tools to be a journalist.

You would need to attract attention from outside your own community, and start attracting a following. This book isn't about that, but there are many articles on the web to give you ideas.

Where traditional journals (and journalists) score is in the credibility. However, research published in April 2011 showed the extent to which the traditional and new media worlds are intertwined. A survey of more than 1,000 business journalists around the world produced the following statistics:

- Social media is the most increasingly influential source of information on stories published by business journalists.
- However, it does not influence journalists as much as their own research or more traditional information sources.
- On balance, social media is seen to have had a positive effect on the quality of the journalist, and will become increasingly important to the angle and content of published stories.
- Nine out of ten journalists claim to have investigated an issue further from social media sources for their work.
- Two-thirds claim to have written a story which originated in social media, giving rise to one in seven published stories.
- Twitter provides the most valuable sources of information, yet blogs are most likely to be the foundation of the published articles.
- Journalists in North America are more likely to use and believe in importance of social media than those elsewhere.[6]

This has several potential effects on you as a scientist, researcher or physician: Your work is more visible than ever, and you are likely to be contacted by people who want to circulate it, comment on it or criticise it. A few simple 'retweets' on Twitter can potentially get your research or summary seen by many thousands of people. Given the statistics above, it is very likely that there will be journalists among them.

Mainstream journalists can find you very easily, and ask you to comment on somebody else's work. To test this out, I just set up a Twitter stream called 'platelet biology' and immediately found more experts than I would ever need. If I need a comment on a story, I could get one very quickly. At this moment, a journalist could be setting up a search filter which leads them to you.

---

6   Brunswick research, 2011. http://www.politico.com/static/PPM153_social.html. Accessed 01.01.12.

The internet is completely unregulated and unchecked. In addition, anyone can set up a professional-looking website or blog, and appear to be a reputable organisation. In reality, this means that although there are a lot of sites and commentators which in reality have little scientific credibility, but have attracted a huge following of like-minded individuals. Check your own specialist area for examples, or look at some of the wilder claims regarding alleged damage from vaccines.

Mainstream Media

Despite all this, if you present your research at a recognised meeting or congress, you are likely to come into contact with traditional journalists. They have different roles and responsibilities, and it will help if you understand them. Before I look at the different types of journalists, a word on one type of organisation you may not have encountered: news agencies.

A news agency is an organisation with its own reporters, correspondents, photographers (and sometimes video crews) which covers events on behalf of its clients, usually news organisations. The main international news agencies are Reuters (based in Europe) and The Associated Press (based in the US). Agence France-Press is the main agency for the Francophone world, and there are others in other languages. Their medical, science, health or pharmaceutical correspondents attend most scientific congresses of any size, so this is where you may meet them. This means that the agency reporter may cover your presentation, or the press conferences about it, and the story which they write is then distributed to the world's media. That's how one report can find its way into more than 1,000 publications around the world. That's a lot of exposure! Given that the premise of this book is that you want to start climbing the professional ladder of recognition, you can see how important the media is.

Most news organisations have a similar structure, with the editor at the top, a deputy editor then a number of assistant editors with responsibility for different areas (home news, foreign news, sport, health, features, pictures and so on). Below that in the hierarchy are a number of types of journalists. For simplicity, I have used newspaper terminology here, but the structure is broadly similar in TV and radio stations. Typically they are:

COLUMNISTS

The high profile voice of the paper. Powerful within the organisation, and usually opinionated. If are doing controversial research, expect them to comment on it, usually without contacting you.

CORRESPONDENTS

Senior journalists with responsibility for their specialist area (but remember my earlier comments about 'specialists'). They advise the editors of the paper on the importance or relevance of stories, and are an important part of the decision-making process for inclusion or exclusion of topics and articles. In your world, you may be contacted by medical, science, technology, pharmaceutical or health correspondents. In some organisations, for example women's magazines, 'health and beauty' or 'health and wellbeing' are bundled together into one subject.

REPORTERS

Divided into two types: General and specialist. General reporters cover any kind of breaking news or diary event, and rarely have much specialist knowledge. They may cover a press conference about new data, but would be unlikely to cover a presentation at a scientific congress.

However, they may be involved in following up a report which has appeared in a specialist journal, and may contact you for further explanation or a comment. As an example, many years ago I was a general reporter on a large regional newspaper in the UK, and part of my job was to read the *British Medical Journal* (*BMJ*) and *The Lancet* every week, looking for stories that would interest our readers. Thankfully, both publications came with a helpful press release attached, aimed at journalists like me who at the time had no specialist knowledge.

OTHER TYPES OF JOURNALISTS

Behind the scenes are many other journalists who are involved in producing the newspaper, programme, website or blog. You are unlikely to come across them, but they have a major influence on whether your story is covered and the angle taken. They include commissioning editors, desk editors, item editors, sub-editors, layout artists, producers, researchers and in some places, fact-checkers. The main point in you knowing about them is to understand that the reporter is only a small cog in a big wheel. They will try and report your story accurately, but if mistakes creep in, it may not necessarily be their fault!

THE JOURNALIST'S WORLD

To make the most of the media, you need to understand how journalists think and work, what drives them and what challenges they face in their profession. You also need to understand what makes a story, and how to make it relevant to their audiences. Helping you to do that is the aim of this section.

The journalist's world can be summarised by these five lines:

1.   Headline
2.   Deadline
3.   Byline
4.   Good line
5.   Bottom line

Headline

At its most basic, the headline is the summary of the story. In reality, it may go beyond that, and may predict what will happen as a result of the event it is reporting. It is designed to be eye-catching, and may be controversial. Unlike your original presentation on which it is based, it will concentrate entirely on the potential benefits of your research to its audience.

Here are some recent headlines, reporting newly reported trials, from the UK and US media:

Once-a-Day Asthma Pill 'Is More Effective Than Inhaler'

Call for All to Take Statins After 55

UK Nuclear Power Plants Cleared of Causing Leukaemia

Breast Cancer Prevention Drug 'Only Works in Half of High Risk Women'

Drug That Stops Bleeding Shows Off-Label Dangers

As you can see, these appear to be fairly factual summaries of the studies concerned. In fact, the first headline is inaccurate. The study did not say that the asthma pill was 'more effective than inhaler'. It said the efficacy was about the same, but that the pill might be better for patients who had problems using inhalers. The text of the story made this clear, but the headline, written by the sub-editor, was incorrect. This is the kind of inaccuracy which creeps into medical stories, based on the news process I explained earlier.

A question people often ask me is: 'How can I prevent being misquoted, or avoid my trial being misreported?' The answer is that you can't prevent it completely … just as all drugs have side effects, every media interview carries some risk. However, you can mitigate the risk by asking the right questions. You can also have a very clear idea of what headline you would like to see on the story, and building a case which is likely to deliver that. Also, think of the headline you don't want to see, and do everything you can during the interview to minimise the chance of that happening.

Deadline

On my first day in journalism I was taken round by an old-style hack who informed me, 'Journalists are experts at dealing with the three Ds … Deadlines, Drink and Divorce.' I will leave others to comment on the other two Ds, and concentrate here on the first one. The deadline is the time by which the story needs to be ready for publication or broadcast. That seems straightforward enough, assuming you are given enough time to finish it. In reality this was rarely the case in news journalism, and over the years the time pressure has increased hugely. Looking at one of my own former employers, Independent Television News (ITN) in London, illustrates how this has happened.

For years ITN, the news supplier to commercial television in the UK, produced three programmes: At lunchtime, early evening and late at night. Each one had its own deadline. Then they added Channel 4 News at 7 p.m. (produced by a different team but with some overlap). Then they added regional headlines every two hours throughout the day. An all-night service came next, called ITN: Into The Night. Finally, after the success of Cable News Network (CNN) and Sky News, ITN launched its own 24 hour news channel.

In a 24 hour news operation, the deadlines merge. The journalist has just finished watching their piece go out on the 1 p.m. news when the producer of the 2 p.m. news wants some changes made. Then the 3 p.m. producer wants something else, and so on throughout the day. The effect of this is that journalists are now constantly on deadline, and under pressure to produce stories. This leaves little time for research or anything other than the most cursory of fact checks.

What this means to you is this: the closer your story is to readiness when you talk to the journalist, the better your chances of success. 'Readiness' here includes a clear idea of what you want to say, preparation before an interview, and an understanding of the journalist's audience and needs. This includes photographs, explanatory graphics and anything else that will aid comprehension. I will return to this topic later in this chapter.

Byline

When I ask people what a byline is, hardly anybody outside journalism and Public Relations knows the answer. That is interesting because to a journalist, it's possibly the most important concept involved in any story!

The byline is the reporter or writer's name on the article. As in, 'By John Clare.' That's it. The reasons behind it are multifactorial: It's a combination of ego, bragging to rivals, rewarding success, building an image and reputation, staking an exclusivity claim and typography. To the reporter, however, it means one thing only: It's a measure of their success in a tough industry. Bylines mean recognition, and bigger bylines mean bigger recognition. Picture bylines (where the author's photo also appears) means you are an important part of that organisation's team ... at least for the present.

Bylines are currency in journalism. I might be introduced to a reporter or correspondent for the first time and say, 'Hi ... good to meet you. I've seen the byline, obviously.'

This is important to you because if you pitch a story to a journalist (or someone does it on your behalf) their first thoughts are 'What's the headline? Will this story get my name in the paper? When do I need to write it?'

Good line

A 'good line' is journalistic slang for an eye-catching phrase, or a good quote which summarises something more complicated. Check the news websites in your country for your own area of science and you will find them. Here are some examples:

> *In the future, predict and prevent will become as important as diagnose and treat.*

> *It is humbling for me and awe inspiring to realise that we have caught the first glimpse of our own instruction book, previously known only to God. (After the announcement of the success of the human genome project.)*

> *Human cloning should not be out of the question. In vitro fertilisation was once seen as depraved God-playing and is now embraced, even by many of the devoutly religious. Cloning could be a blessing for the infertile, who otherwise could not experience biological parenthood.*

A 'good line' will increase your chances of getting publicity for your research. In fact, it's an essential part of any story. If you are really skilled you can deliver it as part of an interview, knowing that it will become the focus of the follow-up story. A great example of this was a professor of epidemiology who was interviewed on BBC Radio 4's flagship morning news programme, *Today*. He had been invited on to comment on a study which had apparently identified an alarmingly high rate of deep vein thrombosis (DVT) on a particular brand of oral contraceptive. This was after previous stories had identified an increased risk of DVT on earlier oral contraceptive pills, so there was strong circumstantial evidence that the new claim could be correct.

The interviewer said, 'Professor … this story is deeply worrying for the millions of women on these kinds of pills. And we can't be sure if the increased thrombosis risk applies to other contraceptive pills, as well. This is a real cause for concern, isn't it?'

The professor replied, 'No it's not really. We should take it seriously of course. But there's no evidence to be concerned at the moment.

> *Interviewer: 'Why ever not?'*

> *Professor: 'It may become a concern, but it's impossible to say at the moment. The researchers have fallen into what I call the Judas trap.'*

Interviewer   (becoming more interested as he senses the controversy building): 'The Judas trap? What do you mean?'

Professor: 'Well, it's about the small sample size. Statistically, Jesus was betrayed by more than eight per cent of his disciples. According to the bible he was actually betrayed by one man ... Judas. You can't extrapolate that to widescale betrayal, and you can't extrapolate this result to millions of women. We need more research.'

The headline on the follow-up story was 'Pill Blood Clot Risk Dismissed as Judas Trap by Professor.' As a widely-quoted scientific academic, he understood exactly what he was doing, and knew precisely when to drop his good line.

Bottom line

This phrase has a number of meanings. It refers to the fact that almost all media organisations need to make money, and deliver bottom line returns for their investors. Journalists rarely think about that, as their aims are simpler: Getting a story and their name in the paper come joint first, and if that can involve some truth-seeking and exposing wrong doing, so much the better.

The bottom line for the journalist is this: 'I'm doing this story, and it's going to run in my paper/programme/website. Whether you choose to talk to me or not is a matter for you. The story is going in anyway.'

I regularly ask scientists to give me a word or phrase that summarises what they believe journalists do in the reporting of science. The most popular answers make interesting reading:

- sensationalised
- over-simplified
- out of context
- selective/report only part of the story
- stretching the facts
- unrealistic extrapolation
- don't understand science
- pre-conceived idea
- biased against science
- look for controversy and conflict
- confuse fact and opinion
- give all views equal weight

Sometimes, different news organisations take completely the opposite views on the same story. Here's an example of two conflicting newspaper headlines reporting the same studies.

Screening All Older Men for Prostate Cancer 'Could Reduce Deaths by a Third'.[7]

Prostate Cancer Screening May Not Reduce Deaths.[8]

Given that list of negatives, and the overall confusion, why would you talk to the media? It does represent a real dilemma for many scientists: Do they refuse to talk to the journalist because of a fear that at least some of those factors may come into play? Or do they talk to the journalist anyway, and try to make the story as accurate as possible, even if the truth is not as sensational as the reporter hoped?

My own view is that in most cases, you should do the interview, as that is the best way to increase your chances of telling the story accurately. I hope the guidelines in this section will help you to feel more equipped.

What Journalists Want in a Health Story

Having outlined the media landscape, I would now like to offer a detailed checklist of how that translates into a wish list, and illustrate briefly how you as a scientist can take advantage of that.

What journalists want in a story is pretty universal, and applies across different subject areas. Here is a list of attributes of a good story, with examples of what this means to your own world of clinical and scientific data:

- different
- relevant
- understandable
- interesting
- supported
- = 'Newsworthy'

7   *Daily Mail*, 19 March 2009.
8   *Washington Post*, 19 March 2009.

DIFFERENT

'News is change' said the US TV news pioneer Reuven Frank. He went on to say, 'News is change as seen by an outsider on behalf of other outsiders.' This quote summarises exactly the first requirement of news, whether it is scientific, financial, political, economic, sports or the weather. It must be reporting something different, saying something that could not have been said before this news broke. In your case, this means before your trial results were made public.

The magnitude of the change must be sufficient to be seen by others, who are not involved in that field. So an incremental improvement in success rates of a medical intervention will probably fail this test, where a significant improvement would pass. The 'outsider' here is the journalist.

The 'other outsiders' are the public, or the audience, on whose behalf the journalist reports. This can be expressed another way in the question, 'What does this mean to me?' If you fail to answer this question the journalist may ask it another way: 'So what?' Avoiding 'So what?' while still remaining true to the data findings can be a major hurdle for scientists who want to engage with the media.

RELEVANT

Journalists have a very clear idea what kind of people make up their audience. They know their predominant political views, income brackets, the kind of cars they drive, the kind of places they go on holiday. They understand their fears, hopes and expectations. One of their major fears, journalists understand, is getting sick. One of their major hopes is being cured. A major expectation is that science will continue to make progress and develop better medicines and procedures. You regularly hear people with long-term illnesses say they hope they can live long enough for 'something to come along that can cure me'.

The kind of stories you will have, based on new data, tap into all of these feelings. That's why health and science is such a good topic. In my lectures I often use a slide which includes a photograph of a TV news presenter in a studio. The caption behind him says, 'Canadian Cancer Cure'. Those three words tell us a huge amount about what journalists want in a science or health story.

CANADIAN

The photograph is from a Canadian Broadcasting Corporation news programme. The fact that Canadian scientists have apparently developed a 'cancer cure' gives the viewers a little bit of ownership of it. At least some of the research was probably funded by taxpayers anyway, so they probably have every right to feel a sense of ownership. The psychological pull, however, is stronger. Canadians are very aware of the huge country (in terms of population) just to the south, so anything that they can justly claim to be their own makes them walk a little taller.

CANCER

Statistically cancer is one of the world's biggest killers. Emotionally it is probably the most frightening disease. The thought of a lingering painful death from cancer is at the top of most people's list of things to avoid. One in three of us will suffer from some form of it in our lives, so we are all aware of its proximity. That grabs our attention. The sheer size of the cancer problem makes it relevant to all of us.

However, there is another reason: If you imagine a curve of distribution, the parts of the curve most favoured by journalists are the extremes ... the very big or small, very long or short, the obese or anorexic. This is because, in the English language, journalists' favourite words end in the letters '... est'. Words like biggest, smallest, richest, poorest, cheapest, dearest, fastest, slowest. Alternatively they like words which end in '... st' such as first, last, least, most. All of these terms are extremes, and this '... est concept' is crucial to understanding what journalists want in a science story. Quite simply, the more '... est' words you can use in your presentation, or in your interviews, the higher your chances of attracting media attention.

Of course this poses another dilemma for scientists: science and medical research advance slowly, and the language used to describe them is usually measured, even conditional. A development might come gradually into focus over years, as different researchers identify new targets and tweak new molecules to attack them. News journalists, on the other hand, imagine a 'Eureka!' moment, when a bald , white-coated elderly scientist runs down a corridor shouting to his colleagues, 'I've done it! I've done it!'

Ironically, a candidate for the greatest scientific achievement of the twentieth century, the discovery of what we now call deoxyribonucleic acid (DNA) by British scientists James Watson and Francis Crick, contains both the scientist's and the media view of how such things happen. In their paper announcing the breakthrough, published in *Nature* on 25 April 1953, the pair, who later received the Nobel Prize for this work, began slowly and modestly, using a slightly different term from the one we use today:

> *We wish to suggest a structure for the salt of deoxyribose nucleic acid (DNA). This structure has novel features which are of considerable biological interest.*

However, two months earlier, having completed their work by 'thinking around the problem' rather than carrying out lots of experiments, they walked into the Eagle pub in Cambridge, and Crick revealed they had just experienced the 'Eureka' moment (a phrase Watson actually used in a later interview) when he said, 'We have discovered the secret of life.' That quote has stuck with him ever since, as an early example of a great media soundbite.

CURE

No word in the English language embodies the feeling of hope more than this one. It is the holy grail for medical researchers, physicians, patients and loved ones. When combined with cancer, it's unbeatable as an enthralling concept. The two words together are so powerful that they are often used to demean something else which is regarded as cheap or unimpressive. In this sense people will say, 'Well it's not a cure for cancer, is it?' In addition, 'cure' also embodies the concept of extreme discussed earlier: On the distribution curve I mentioned, 'cure' would be at one end and 'dead' at the other.

UNDERSTANDABLE

Journalism: A profession whose job it is to explain to others what it personally does not understand.[9] Much of the data you present has the potential to be made relevant and different. The real challenge is to make it understandable, first to journalists then to their audiences. This is where many scientists fail, predominantly for two reasons:

---

9   Lord Northcliffe, British press owner.

1. They overestimate the understanding of the journalist.
2. They regard making something simple as 'dumbing down', and fear the criticism of their colleagues.

There are two main ways in which a story can be made difficult to understand: Difficult concepts and difficult language. In science, it is impossible to avoid difficult concepts. Your responsibility, as a presenter or an interviewee, is to explain them in words that the audience understands. That's where the right language comes in. The English poet Samuel Taylor Coleridge (1772–1834) said:

Prose is words in the best order

Poetry is the best words in the best order

My advice here is simple: be a poet!

Being clear and understandable of course is not just a requirement when talking to the media. I hope I have made its importance clear throughout the book. My mantra, which is printed on most of the materials my company produces, is:

*Good communication starts with saying something
in a way that can be understood.*

*Great Communication starts with saying something
in a way that cannot be misunderstood.*

Aim high: Strive to be a great communicator!

To see great communicators in action, check out top scientists such as Richard Dawkins, Robert Winston or Brian Cox. All three of them are great communicators.

INTERESTING

This is the reason we buy newspapers, watch TV news and current affairs, log on to our favourite websites … because what we find there is interesting. Journalists have very finely tuned antennae for what is interesting (and what's not) to their readers. They are also very skilled at creating an interesting angle if they think it will work. The factors above (relevant, different and understandable) increase the likelihood of something being interesting, but some subjects are only interesting to a narrow group of experts. As an example, earlier this week a client sent me a paper which had been published in a speciality journal. It was the results of a study comparing clinical trials with clinical practice. Personally, I find this subject very interesting as the challenges of extrapolating trial results to the general population of patients can be great.

I think the 'trials v practice' concept can be made interesting with the right examples. Patients sign up for trials because they get better care, regular visits to the physician, they're made to feel special, compliance is high, and they may get early access to a new medicine that could really help (or even cure) them. Most of those factors do not apply in day to day clinical practice. So there is huge potential for an interesting story, relevant to many people, which could attract the attention of the media.

However, this particular paper focused on one aspect of trials v practice: In trials with a switchover design, a washout period is often built in, to ensure the 'old' drug has been removed from the patient's system before the 'new' drug kicks in. In clinical practice, if patients don't tolerate one medication they are usually switched to another, with no washout period. The question asked was; 'Do we get different results if we conduct trials without a washout period?' The answer was complex, but boiled down to 'sometimes'.

Keen as I am on the overall topic, my view was that this study was of limited interest to a small number of people, and had virtually no chance of attracting attention from mainstream media. The data was being presented at a satellite meeting at a conference on pain, and would probably be well-received by people who were very involved in that world. However, as a media story, it didn't make the cut. It's also a good example of the Reuven Frank quote I used earlier, 'News is change as seen by an outsider on behalf of other outsiders.' The magnitude of change here was too small to be noticed by any outsider.

SUPPORTED

Journalists are sceptical. A crucial part of the journalist's character is to question statements, and seek other views. So, if the Principle Investigator (PI) of a study describes it as a landmark trial, the reporter will ask someone else to corroborate that view. You can pass the scepticism test by using the Assertion – Evidence – Support formula for your media statements, as outlined in Chapter 3.

In another sense, if you are the PI on a big trial which shows drug A to be safer or better than drug B, you are the support for the pharmaceutical company's claims about it. Understanding your role in a story is a crucial part of successful media handling.

NEWSWORTHY

If your story fulfils all the criteria outlined in this section, it becomes what journalists call 'newsworthy', that is, 'worth including in the news'. If it fails, it may be dismissed as 'worthy', that is, 'possibly important, but dull.' Successful media handling means you've avoided the 'worthy trap'.

Chapter Summary

- The agendas of journalists and scientists are often in conflict: Much journalism is presented as black and white, while much science is less definite.
- In the media, fact and opinion are often combined.
- Emotion plays a big role in some media stories.
- Never underestimate their intelligence, always underestimate their understanding.
- Reporters can be generalists or specialists.
- Specialist correspondents do not usually have the depth of knowledge of a scientific specialist.
- The internet has changed journalism fundamentally.
- Social and mainstream media feed off one another.

The journalist's world is composed of:

- headline
- deadline
- byline
- good line
- bottom line

Memorable phrases will increase your chances of media success. Journalists want science stories to be:

- different
- relevant
- understandable
- interesting
- supported

If you achieve that, your story is 'newsworthy'. If not, journalists may regard it as 'worthy', that is, important but too dull for inclusion.

7

Media Interview Techniques

Overcoming media challenges – definition of an interview – off the record – attracting media attention – preparing for an interview – questions to ask a journalist – example of a confusing interview – the importance of addressing the question – bridging techniques.

Interviews Are Not Always Obvious

This chapter introduces you to the world of the media interview. Facing the media offers particular challenges for scientists, as illustrated in the previous chapter. This chapter focuses on how to overcome them. It looks at ways of bringing your research to the attention of journalists, handling different types of interviews, and techniques for spotting when things are about to go wrong.

First, I want to clarify the notion of 'a media interview'. You may imagine it to be a self-contained event, where the journalist comes to your office or clinic at a pre-arranged time, asks you a list of questions, then leaves. This does happen, but it isn't always so formal. In fact, journalists' aim is to find out information, which they can turn into a story. That process may involve 'an interview' of the type described here. However, it may consist of a number of informal telephone chats or email exchanges with scientists they already know, and a couple of minutes on the phone with you where they want the answer to one question and a memorable soundbite. Considering how quickly journalists have to work, and the number of subjects they work on at once, the whole process can seem chaotic to the interviewee. You need to remain focused on your story.

When I teach media skills to scientists and physicians I always begin by asking about their media experience. It's not uncommon for someone to say, 'No, I've never been interviewed' then later in the session talk about a news story they took part in, or an occasion when they were quoted (or misquoted) by a journalist. When I say, 'I thought you said you hadn't been interviewed', they usually reply, 'No, I wasn't interviewed; I just talked to a reporter.' This is a common mistake.

> *If you talk to a reporter, you are being interviewed, and they can quote what you say. You need to be alert, and ready with your story and quotes.*

OFF THE RECORD

Another topic which causes confusion is the concept of 'off the record'. It's an idea more common in political and corporate circles than science and academia, but it is worth understanding the ground rules. Therein lies the problem: Most people (outside the media) don't understand those rules. Here is a situation to test your own understanding:

> *Imagine that you tell a* Daily Gazette *journalist 'off the record' that an Oscar-winning actor and rock star whose mother died of Alzheimer's is donating 50 million dollars to Alzheimer's research at your institution, and the government has promised to match it. This is all top secret, and will be announced next week once the small details have been agreed. The journalist has been very helpful to you in the past, and you want them to know first. The question is this: What can the journalist do with that information, while still obeying the 'off the record' convention? I regularly ask this question in our training courses, and receive widely differing answers. Here is the situation: They can report it, and quote you directly, but not identify you as the source.*

Now the journalist has a good exclusive story:

> *Alzheimer's research has received a $100 million dollar boost from rock and film legend John Smith. He's given $50 million of his own fortune, and promised more if the research is successful. The government, after long discussions with ABC University who will lead the research, has agreed to double the donation....'*

So far, so good. Now let's look at the next part of the story, which involves you.

The Gazette learned the story exclusively last night from a source at the very top of the university's Alzheimer's research team. I was told, 'This is just the shot in the arm we needed to fund the next part of our research into the genetic triggers for this devastating illness which is destroying more and more lives as the population ages. $50 million will make a significant difference ... we're very grateful to John. It's been a hard task getting the government to match the money, but to give them their credit, they've now finally coming through with the cash.'

The convention of 'off the record' means you cannot be identified. So how do you feel about being referred to as 'a source at the very top of the university's Alzheimer's research team.'? The journalist wants to position their source as being senior and authoritative, so that seems fair enough. Let's take that one step further. Imagine that the secret negotiations with the government were conducted by a committee of four people, including you. Would you be happy to be identified in the story as a member of that committee? What if there were two male and two female members, and the story referred to you as 'she' later in the text? This actually happened in an exclusive story about reorganising the NHS in the UK, when the source was accidentally identified as female, ironically by a female journalist.

You can see the difficulty here: Without identifying you by name, the journalist can give some very good clues as to who you are. But there is another problem: Asking to go 'off the record' can sometimes send out the wrong signals. They may ask themselves, 'Why can't they tell me this on the record? Why the secrecy?' It can make them suspicious, so you need to allay their suspicions. Sometimes the reasons may be obvious, but sometimes you need to spell them out. Honesty may be the best policy.

To illustrate a more worthy use of off the record, imagine another situation: You have discovered that cutbacks in your hospital funding have led to a reduction in cleaning, with an increase in potentially fatal MRSA and *clostridium difficile* infections. You regard this as a real thread to patients' lives, and think that the hospital management needs to be shamed into restoring the cleaning budget by the facts being exposed in the media. Employment contracts for many hospitals and other public institutions now include non-disclosure clauses forbidding you to speak to the media without the institution's consent. Journalists often regard these clauses as 'gagging orders' and a threat to freedom of speech. Anybody who disobeys the 'gagging order' to leak information which is a legitimate public concern becomes a 'whistleblower', and is generally welcomed by journalists.

If you are subject to a non-disclosure clause and want to bring the increased infection rates to the media's attention, you can be honest about that. You can say to the journalist, 'I'm not allowed to speak to you under the terms of my contract, but there is something happening which I think should be publicised. However, first I need your word that this off the record conversation will remain off the record. I can give you specific figures which it will be impossible for the management to deny, but that will provoke a 'mole hunt' within the hospital. I can tell you how to verify the information, and who to speak to. But this must not come back to me in any way, can you guarantee that?'

You can see that such negotiations are fraught with difficulty. That's why 'off the record' should only be used by experienced media handlers who know how to use it.

Attracting the Media

I am now going to turn briefly to more traditional (and less risky) ways of attracting media attention to help you publicise your research. This section is intended as no more than an outline of general principles, because most institutions today have their own Press Offices and Media Relations team, who will do most of the outreach on your behalf. When I refer to the media in this section, I mean the general media, not the specialist peer-reviewed journals where your paper would ideally be published. Involving them is an entirely different process outside the scope of this book.

The first step is to get a realistic evaluation of the kind of media that might be interested in your story. Options include (in order of difficulty):

- National press, TV and radio.
- General interest magazines (for example, *Cosmopolitan, Bella, Marie Claire* in the US or *GQ*, other men's magazines.
- Regional media (newspapers, TV, radio).
- General magazines focused on health, for example, *Top Santé, Men's Health*.

- Magazines aimed at healthcare professionals, for example, *GP*, *Pulse, Hospital Doctor* in the UK. [BMA News?]
- Websites, either independent or linked to any of the above.

The next step is to understand the news cycle, or the production routine of your media target. This starts with a weekly planning meeting and ends with publication. You need to ensure that a summary of your story is available for consideration by the editors at the planning meeting. The timing of this meeting varies, but on a daily paper or TV news programme, it is often held on a Thursday. At the meeting, the heads of the sections (home news, foreign news, features, entertainment, sport and so on) present their ideas for stories for the following week to the editor. Health and science are normally sub-sets of the home news (or features) sections, so would be presented by those section heads. The advantage of having your story considered at this meeting is that it can receive (or fail to receive) 'buy-in' from all the important executives on the paper. Once it has the stamp of approval, any programme or section editor knows they can include it without any further discussions with the bosses. If your story doesn't make this meeting, or is rejected, it has a more difficult passage through the production process.

That leads to another key question: how do you get it onto the planning list? There are several ways, but none of them trumps personal contact. If you (or your PR team) know the health/science/technology correspondent, the best way is to contact them with a summary of the story, the timing and brief details of potential interviewees (that is, you!) If you don't have that kind of personal contact, you can contact the relevant correspondent anyway. Be brief and to the point, and expect to be talking to someone who is extremely busy and may be doing other things while listening to you. In fact, you may even find that you are talking to someone else in their office while you are trying to sell your story … it can be intimidating, but persevere. Another way is to contact the planning editor, sometimes called the forward planning editor, and pitch your story there.

Pitching your story to journalists is a specialised task. That's why it is normally done by people who do it all the time. In the case of hospitals and academic institutions, this normally means the PR department. In pharmaceutical companies, it may be done by the in-house team or may be the responsibility of the PR consultancy. The advantage of the consultancy is that they pitch stories all the time, and will have good relationships with the key correspondents.

Questions to Ask

The process outlined above relates to *proactive* media handling, that is, when you (or someone on your behalf) approaches the media. Frequently, however, it happens the other way round, and a journalist will approach you. We call this *reactive* media handling. It may happen, for example if the journal where your research is published has sent out its own media release, or their own media team have used their own media contacts. It is important to realise that even the tier one journals compete for status, circulation and reputation. If they have a good story, they want to see it in the general media, with a credit to the journal.

If you are approached by a journalist, here are the key questions you need answered before you decide whether to do the interview. Ideally, you would ask someone else to find out the answers for you before you talk to the journalist yourself. Your PR team if you have one, will certainly do this as a matter of routine.

WHO ARE YOU AND WHO DO YOU REPRESENT?

Many journalists work for huge organisations, for example, The BBC or ABC TV, or News Corporation (owners of *The Sun* and *The Times* in the UK and *New York Daily News* in the US, which are all aimed at very different audiences). Others work for news agencies, as outlined in the previous chapter. You need to know precisely where the interview will be used, because that enables you to frame your answers in a way which is relevant to the audience. If you haven't heard of the publication they are working for, ask them! This is particularly important with websites. Ask for the link, and take a look at it yourself to make sure you are pitching your answers at the right level of complexity, and making the benefits to that audience completely clear.

The journalist may be a freelance, and within that world there are many different types. They may write for a number of different outlets, or be a regular contributor to leading newspapers and magazines. Others are just on a 'fishing expedition', hoping to find a story they can sell to somebody. Talking to them may not be a good use of your time.

The other information here concerns the reporter's speciality. I outlined the key ones (health, science, medicine, technology, finance, pharmaceuticals) in the previous chapter. You need to tailor your story to fit the reporter's agenda, audience and level of understanding.

WHAT DO YOU WANT TO TALK ABOUT?

There are several levels of answer to this. At the top level is the basic topic. In your case, it's probably clear that they want to ask about your research. In large institutions and corporations, however, there are so many potential stories at any one time that you need to ensure that you know which one the journalist is following.

The next level of answer concerns the detail of your research. It is quite likely that the journalist wants to ask you what the consequences are, whether it means a cure may be possible, how long before X treatment becomes available, or whether you feel there are particular types of patient who may be helped by it. None of these questions should pose any difficulty for you, but you need to prepare your answers.

Another level of enquiry may be using your research as an example of a wider picture. For example if the government is cutting science funding, your study may be the kind of work that would not be funded under the new arrangements. If you're unlucky, it might be held up as an example of the kind of blue sky research which has little practical application, and would not be missed if it did not continue. Only you can decide whether to get involved in interviews of these types.

A well-known professor in the UK was recently asked on national radio about the point of what the interviewer called 'useless research'. Her answer was, 'It isn't useless research, it's research for which society has not yet found a use', which I thought was another excellent soundbite.

She also had a point: I once interviewed Arvid Carlsson, after he won the Nobel Prize for Medicine in 2000. He won it for his work on the identification of dopamine as a neurotransmitter in the brain. He made his discovery in 1957, but its relevance only became apparent years later, with the development of anti-psychotic drugs known as selective seratonin reuptake inhibitors (SSRIs). The 43 years between his discovery and the awarding of the prize was the longest in Nobel history at the time.

HOW MUCH DO YOU KNOW ABOUT THIS TOPIC?

Medical and science journalists come from a variety of backgrounds, as discussed in the previous chapter. In addition, once your story gets out into the mainstream general media, you may be interviewed by general news reporters. You need to establish how much they know about your topic, but do so in a way without alienating or insulting them, or making them feel small. A few non-threatening questions or comments will help, such as, 'Have you written much on this topic?' or 'I'm not sure how much you know about this, but let me sketch out the basic idea ...' Your own social skills, and the questions the journalist asks in response, will tell you whether you've started too simplistically, or are getting too complex.

HOW LONG WILL THE INTERVIEW BE?

This relates to two separate matters:

1.   How much time do you need with me? (Face to face or on the phone)
2.   That is, how much time should I allow for it?

HOW LONG WILL THE FINISHED ARTICLE/PIECE PROGRAMME BE?

Is it going to be a 3,000 word article in *Time* magazine, a 15-second soundbite on a local radio, or a significant segment in a 60-minute documentary? The amount of time needed to prepare for these different scenarios varies hugely.

FOR BROADCAST: WILL IT BE RECORDED OR LIVE?

In a recorded interview, you can ask to 'retake' a particular answer if you are not happy with it or think you can make the point more succinctly. The reporter will usually help you with this, because they want short self-contained memorable answers too.

   In a live interview, obviously this is not possible. However, bear in mind that live interviews are usually recorded as well, and a clip is used in later news bulletins. This means that even in a live interview lasting, say three minutes, you should also aim to use short soundbites for later use.

WHO ELSE ARE YOU INTERVIEWING?

It is rare for journalists to write or broadcast a story without talking to several sources. The fact that your research has been published in a peer-reviewed journal is a huge testament to its credibility, but journalists usually like to check it out with their own contacts. Don't be afraid to ask about this. The answers will give you a good idea of the overall tenor of the piece. For example if you've been involved in a trial of a new vaccine, and the reporter tells you they are also interviewing the anti-vaccine group in your country, this tells you that the resulting article/programme will have at least an element of controversy. If they're interviewing a patient support or the chair of the government's vaccination schedule committee, the piece will be very different.

WHAT PART DO I PLAY IN THE STORY?

For example, is it a long piece about this therapy area and you just need a quote from me about the potential of our research? Do you want me to comment on research done by someone else? Is it about our institution's research focus or company pipeline? Is it primarily about patients, investors or scientists, or about the potential impact of a new drug on physicians' treatment options? Where do I fit into this picture? Am I central or tangential to it? Will I be named or just a spokesperson?

WHEN IS YOUR DEADLINE?

See the discussion about deadlines earlier in this chapter.

Having ascertained answers to all the above, you are then in a position to ask yourself the most important questions of all:

- Do I want to do this interview?
- Am I the right person?

If the answer is 'yes' you then need to ask:

- What do I want to say?
- How can I make sure I get my points across?

The next part of this chapter addresses these two questions.

PREPARING AND PRACTISING YOUR STORY

Earlier in the book I outlined techniques for clarifying your story, and making it relevant to the audience. All of these exercises are very effective when preparing for a media interview. The key ones are:

- The elevator pitch
- Features and benefits
- MALES
- Point – Example – Point

In addition, here is a list of key points to increase your chances of success when preparing to be interviewed about your story:

- watch the programme/read the publication/check the website
- understand the audience
- stress benefits, not features
- use non-specialist language
- don't over-claim, or under-claim
- show enthusiastic realism
- offer case studies where relevant
- expect scepticism

Taking all that in mind, practising telling your story is the essential final step. Ideally you should do this as part of a media training session, where your performance can be recorded on video, critiqued and improved. You need to become accustomed to telling the story, then answering questions on it. If this is not possible, you should at least practise it with colleagues. Becoming verbally fluent in your story is a key step to success.

SETTING A BAD EXAMPLE

Here is an example of what language not to use in an interview aimed at the general public, even the intelligent, intellectually curious ones who listen to the BBC's *Today* programme on Radio 4. The headline was that scientists had discovered that depression has a genetic link. This exchange took place:

> Scientist: *We found that specific regional chromosome 3, called 3p25-26 showed strong evidence for what we call linkage.*

Interviewer: So that chromosome was present in the majority of people who had depression down the generations?

Scientist: In our study we found that just about half the families were contributing to the signal that we observed. We found that about 5 per cent of families were strongly linked to the region.

Interviewer: But many others weren't. So what you're saying is that there could be a genetic factor in some cases of depression but not all?

Scientist: Precisely. We think that genetic variations in this region are important in some individuals in some families.

The interviewer then asked the classic question in these interviews: When will it lead to a treatment ? The answer was 'At least 10 or 15 years, possibly' and you could imagine hundreds of thousands of people who were initially confused by the reference to 'the region' thinking, 'So what?' The problem was that the scientist hadn't considered the audience's knowledge, the language he should use, or the needs of the journalist. The scientist should have prepared for the interview with a media and messaging session, as a result of which he might have said something like this:

We have found that some people with depression have the same genetic variant, a genetic abnormality. We have thought this for some time, but now we know what specific variation they have. We know it's on a particular chromosome, chromosome 3. We know that because we studied 800 families where some family members had depression, and found a strong link to this genetic change. It doesn't mean that everyone with the change will become depressed, but it is a risk factor.

It means that we and other scientists can now be more focused on our research because we know where to look. Eventually it may lead to more treatments for depression, though that is a long way off.

Interview Techniques

People sometimes misunderstand the idea of 'interview techniques'. This is because in many countries, there is a proportion of spokespeople who try to treat an interview as free advertising. They expect to go on air and make their points irrespective of the questions. Politicians are the worst culprits, and personally I regularly find myself shouting 'Just answer the question!' at the radio news during political interviews. That kind of interview rarely does anyone any good. The listeners can tell that the politician is wriggling and refusing the answer the question, the journalist gets more frustrated, and the reputation of politicians in general, and that one in particular, takes another turn down the downward spiral.

To make my position clear, here: I am not suggesting that you behave like that. However, neither should you go on and dutifully answer all the questions without getting your points across. The ideal situation involves steering middle course: You have to answer questions, and some serious matters, for example, safety and cost, must be addressed comprehensively, but you should also be able to make your own case. If you understand the journalist's agenda, and have followed the advice on preparation in this and the previous chapter, you should be able to achieve this.

One point to consider here: I referred to TV and radio above, because it is only in those situations where you speak directly to the audience. In other environments, notably newspapers and magazines, the journalist interviews you then reports your comments second hand. That makes it much more difficult to deploy the 'politician's answer' technique.

The techniques I outline here are useful for all kinds of interviews, whether broadcast, print or online. There are specific factors to consider with some media, and I will outline those as appropriate.

TECHNIQUE 1: LISTEN TO THE QUESTION

By this, I mean 'listen to all the question'. It is not uncommon for inexperienced interviewees, particularly in a stressful situation such as a TV or radio studio, either to listen to just the start of the question, then begin formulating their answer. This may work well, but if you listen to the end you may find an easier point to deal with, or another question which makes it easier to get to your key messages.

It's also important to realise that many questions are not as difficult as they seem. Inexperienced interviewees sometimes end up with the rabbit in the car headlights syndrome, where they are incapable of taking any action at all. Sometime people go into an interview fixated on what they see as the weakness in the study. I have had researchers say things like, 'OK … I think I'm fine now, as long as they don't ask about the reason for the combined primary endpoint.' This was after we had spent half a morning honing the messages regarding the combined primary endpoint, and practising answering questions on it! Go in with confidence, and fall back on the practice

TECHNIQUE 2: POINT – EVIDENCE/EXAMPLE – POINT

I outlined this technique in Chapter 3, 'Preparing Your Talk'. When it is used in the context of a media interview I call it PEP, as in 'Pepping up your story'. It enables you to repeat your key message but change the illustration so it doesn't sound repetitive. It also gives more support to your main point.

TECHNIQUE 3: JUMPING TO CONCLUSIONS

Of all the potential pitfalls awaiting scientists in a media interview, this is the most common one. The reason is that it goes against the way they are trained. Scientists want to build a case to a point. Journalists want the point first. Here is how it works:

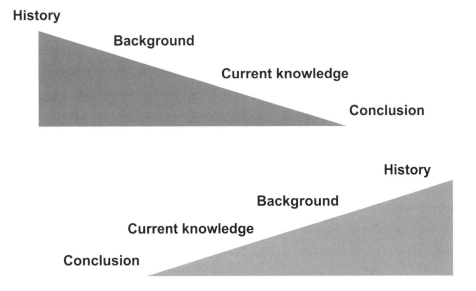

Figure 7.1   Jumping to conclusions

The top triangle illustrates the way doctors and scientists often tell a story. They like to start with the history, fill in on the background, update the audience on the current situation then get to the point. That's why the IMRAD way of writing a trial paper is so common ... Introduction, Methods, Results and Discussion. Unfortunately, journalists like it exactly the other way round. They want the conclusion first, then the supporting facts, and finally the history. There are a number of potential problems with using the top triangle as the template for a media interview:

- It takes too long. You run the risk of being interrupted before you get to the point.
- You are giving the journalist too much information, and many other opportunities to take the interview off in another direction.
- You are not controlling the flow of the interview in a way that suggests the next question. Rather, you are sitting back and inviting the journalist to pick up on anything you have said.

If you use the bottom triangle as your model, on the other hand, you are getting the most important point in straight away. If you are then interrupted, it is less of a problem. You're also giving the journalist less opportunity to take the interview off in another direction.

There is another reason why the 'Jumping to conclusions' technique is so powerful with journalists: It follows the way they write stories (and the way your PR team work with press releases). A news story is usually written to this format:

They start with the intro, which is a summary of the whole story. The main facts come next, giving a little more information. This is followed by the details, and the loose ends tie up any unanswered questions. Almost all news stories follow this format ... take a look at any newspaper and you will see it illustrated.

A benefit of this story shape is that it can be cut from the bottom to cut the available space on the page, and still make sense. In fact, the top paragraph, or the 'intro' often runs on its own in one part of the paper as a 'teaser' for the whole story on another.

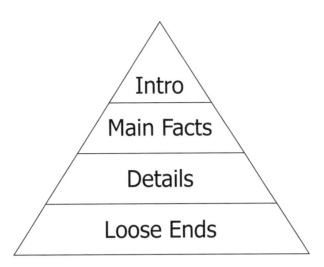

Figure 7.2   News story structure

Given how busy journalists are, and how important it is that you direct their attention to the relevant part of the story, I hope you can see how important the 'Jumping to conclusions' technique is.

TECHNIQUE 4: BRIDGING

Armed with this technique, you can really become the master of the media interview in most situations. However, it needs careful use, quick thinking, and an intuitive sense of when it is (and is not) appropriate. It also needs regular practice for you to use it smoothly. On the downside, when overused or used incorrectly it is the single biggest reason for people believing that media interviewees don't answer the question but go on air with a pre-prepared script which they are determined to get through, irrespective of what they are asked about. As I made clear earlier, that is not the point of a media interview.

The idea behind 'bridging' is simple: You take a word, phrase or idea from the question, and use that to build a verbal bridge to something you want to say. As an example, imagine that you have been invited into a national radio station to talk about a promising new drug on which you have conducted trials. The studies have shown it is safe and effective, but there has been a lot of publicity about the cost of new medications and the effect on the health budget.

Here is an example of how a skilled interviewee would use the bridging technique to tell their story in this scenario, while avoiding the 'politician's answer':

Interviewer: But these new drugs are very expensive, so however good they are, doctors won't prescribe them and the system won't pay for them. That's right, isn't it?

Interviewee: Well the new drugs are more expensive than the older ones. However, as we saw in the trials, the side effects are much less common, and less serious, and that leads to more people taking them as they should, and continuing to take their medicine. So that's important for all of us taxpayers who have to pay for all drugs, not just the new ones. Also, the evidence shows very clearly that they keep people out of hospital, which as we know is the most expensive kind of healthcare.

Interviewer: But the independent think tank the Institute for Financial Studies says that if these drugs were prescribed to everyone who was eligible, it would almost bankrupt the health system. That's unsustainable, isn't it?

Interviewee: Yes I saw that report too, and like you I was initially alarmed. But with great respect to them, I don't agree for a number of reasons. In particular, I don't think they have taken into account the strict eligibility criteria for these new treatments. They will only be available to patients who have failed on the older drugs, initially at least. So if everyone who was eligible for the new medicines got them, there would be an increase in the drug budget, certainly ... but that would be more than offset by savings in the hospital budget. So in my view, patients would have better treatments, doctors would find there was more they could do for those patients, and the people who pay for the drugs – you and me as taxpayers, or as insurance policyholders – should get a better deal.

Interviewer: Well that sounds like a very rosy picture. In reality, though, even in your trials, the drugs didn't work in up to 30 per cent of cases. Surely that's not good for them, to be taking a medicine that might give them side effects but no benefit, and once again it's a waste of money?

Interviewee: Very few drugs work in 100 per cent of patients ... even contraceptive pills don't quite reach that figure, though they're pretty close. The science is advancing all the time, and we are moving towards what we call personalised medicine, which means we can predict which drugs will work in which patients. That will be great when we get there, but for now, 70 per cent is a pretty good figure and compares very well with other treatments. However, one benefit we saw in the trial was that it became pretty clear pretty quickly if a drug wasn't working in an individual. That's what you want, if a drug is going to fail, it should fail fast, then we can try something else, and stop spending that money on something that's not working.

I invented the exchange above to illustrate the point that you need to balance the bridging with answering the question. I think the interviewee did a great job. The most effective way of using the bridging technique is to develop some phrases which are specific to your story, which will help you make a smooth transition. Here are some all-purpose bridging phrases:

- The most important thing to understand here is ...
- You say that, but what people (patients) tell me is ...
- If I can just explain the main point here ...
- Let's not forget that ...
- That's not the case ...
- But equally important, let me say ...
- Let's take a closer look at ...
- That's an important point because ...
- That's an interesting question: However, let me remind you that ...
- Let me answer that by saying this ...
- And don't forget ...
- Before we get off the subject, let me just add ...

However, if you only use phrases like these, you will sound evasive. Before you use them you need to do something really important: Acknowledge the question. That's one of the things the interviewee above did successfully. The bridging technique has three stages, which we call ABC: Acknowledge – Bridge – Communicate. Here are some examples of ABC:

Acknowledge	Bridge	Communicate
Yes that's right …	but equally important is …	key message
That's not quite right …	let me explain …	key message
I'm not aware of that …	what I do know is …	key message
That's a fair point …	but it doesn't change the fact that …	key message
Before we leave that …	I'd just like to say …	key message

Above all, the bridging technique requires practice to sound fluent and avoid appearing evasive. Also remember that you cannot expect to bridge back to your point in every answer, and there are some important issues (for example, safety) that you must address head-on.

Chapter Summary

A media interview can be formal or informal, and may be nothing more than a brief casual chat on the phone or email exchanges

The journalist may gather information from other people, and just want a quote from you.

- 'Off the record' is fraught with danger, and should only be used by experienced media spokespeople.
- Discuss potential story ideas with your PR department, if you have one.
- Alert journalists well in advance if you think you have a story for a specific date, such as the publication of research.
- Ask the journalist questions before you decide whether to agree to be interviewed. Questions include:
  - Who are you and who do you represent?
  - What do you want to talk about?
  - How much do you know about this topic?
  - How long will the interview be?
  - For broadcast: Will it be recorded or live?
  - Who else are you interviewing?
  - What part do I play in the story?
  - When is your deadline?
- Develop memorable sound bites and summary points before the interview.
- Practise the interview with a media trainer or colleague.

- Listen to the question.
- Use the Point – Example – Point technique to bring in examples.
- Start with the conclusion then give the supporting information, not the other way round.
- Use the bridging technique where possible, but avoid sounding evasive.

8

Every Interaction Counts

Informal talks – being ready to talk – the power of a conversation – spoken versus written language – portraying authority, approachability and passion – putting it all together – key techniques – a personal post-script.

The Overlapping Rings of Communication

In Chapter 1 I introduced you to the overlapping rings of science communication, illustrating how the publication, presentation and interviews all combine to produce an overall picture of a development in science, medicine or health. In reality, the three rings are themselves inside a bigger circle. The bigger circle represents all the conversations you have which do not fall into one of the three main categories.

They include the chance meetings at the water cooler, in the coffee queue, on aeroplanes, in the elevator, at a congress or an industry event. They also include scheduled meetings with colleagues and other influencers where you have the opportunity to explain your research or tell your story but do not make a formal presentation.

When you have a story to tell or want to influence people, you have to make every interaction count. Some people are very good at this. Early yesterday morning I arrived at London's Heathrow airport from a business trip to Philadelphia. As we were waiting for the aircraft doors to open I heard a fellow passenger ask a flight attendant about the chance of making a connecting flight to Paris which was due to leave in just 45 minutes.

A short time later I found myself on the transit train between the gate and the arrivals lounge, standing next to the woman who was going on to Paris. I asked if she was going to Paris for a weekend away and she said, 'Oh, no … I work for The Alzheimer's Society. I'm going to the European Dementia

Federation meeting. There's lots of really good research going on in the area, which is really important because there's going to be a million people with dementia in the UK in eight years' time. We need money, and in Paris I'll see great examples of where it's being spent.' I congratulated her on her pithy pitch and clear message. 'Well', she said, 'I reckon you've got to tell everyone about this because it might hit any of us … we're all getting older and we'll need effective treatments.'

It was a great example of 'Every interaction counts'. She was ready with her key message, delivered in a pleasant, conversational style with a smile to a stranger on a train at 6.30 a.m. on a Saturday. She had no idea who I was, or that her message would end up in this book, but she was ready. To illustrate that I practise what I preach, I also took the opportunity to tell her what I do, and promised to connect with her on a social networking site.

I saw another great example when I turned up at a bar in the UK to watch a big soccer game with a group of friends. I introduced one of them, a senior executive in the oncology business at a large pharmaceutical company, to another, who runs an IT consultancy. 'Oh … cancer drugs … they're very expensive, aren't they?' said Mr IT. Mr Pharma hit it back to him, straight off the bat, 'They are expensive because research is expensive. This year we'll put eight billion US dollars into finding new medicines, and hundreds of millions more making sure that people in financial difficulties get help to pay for them.' Delivered with a sympathetic face and a shrug, it was an entirely appropriate response to a legitimate question. As with Ms Alzheimer's, Mr Pharma was ready.

The two examples here worked because the people involved were ready to tell their stories. Their elevator speeches were primed and ready to go. I imagine people like this as behaving like gunslingers in the Wild West, but constantly ready to shoot with messages and stories rather than Colt .45s. They also produced appropriate responses. They didn't lecture, or pull out pie charts and bulleted lists. They told their stories briefly in a conversational fashion. That is what I want to turn to now, and pull the main strands of the book together.

The Power of a Conversation

When I am running story telling or message development meetings with clients, there is usually a moment when I say, 'OK let me try and tell your story for you. As I understand it, it's this …' I then tell them the story as we have developed it so far. The participants often have their own suggestions for improvement, and over time we may change it completely. However, one comment that is sometimes made is, 'That sounds great. Can you do the interviews for us?' I reply, 'That's very kind of you, but let me ask you a question: Why do you say it sounds great? What is it about the way I say it that works for you? I ask you because you can do it too … we can work it out together, now, in this room. You can tell this story as succinctly, powerfully and memorably as I do. But you have to tell it in your own way.' The key here is that it needs to sound like part of a conversation rather than a message track.

Sounding conversational is a really powerful weapon when you want to communicate about science, health and medicine. It has a number of benefits:

First, if you can tell your story naturally without any visual aids, you come across as confident in your topic. You project yourself as an authority, so people are more likely to listen to you. Think about what happens in a consultation between a physician and a patient. If the physician talks authoritatively and conversationally about the condition and gives a confident diagnosis, the patient is more likely to be convinced and reassured than if the physician has to read from a book or computer screen to tell them what's wrong.

Second, being conversational means engaging with the other person in the conversation. They focus on you and what you're saying, and you do the same to them. It produces a genuine dialogue.

Third, it makes you appear more approachable and prepared to answer questions.

Fourth, remember what I said in Chapter 5 about using visual aids? You are your most powerful visual aid. You can make your story memorable by the way you tell it. Think of a stand-up comedian and why they are memorable. It's because of the way they tell their stories. Fortunately, you don't need that level of conversation skill to be an effective communicator about science!

Being Conversational

Occasionally I meet someone who is a very good communicator when they are speaking personally, but not so good when they are talking about business or other professional topics. Some time ago I was helping a senior pharmaceutical executive to prepare for an interview on an investigative TV documentary.

Under pressure, his demeanour became tense, his voice strained and his manner of speaking clipped and formal. To illustrate how I wanted him to speak, and how persuasive he was when he was 'being himself' I secretly recorded him during the coffee break, when he and I were discussing cricket. I showed him the video clip and he was amazed at the difference. Listening to him talk about cricket was a pleasure. He conveyed authority, approachability and passion. Talking about the business he sounded like a different person. It was as though he'd learned some long legal statements by heart, and was not convinced by them. As a result he was unconvincing.

So how do you portray authority, approachability and passion? It begins with a thorough understanding of the topic. I am going to assume here that your knowledge of the subject is not in dispute. What is under scrutiny is the way you communicate it. You need to follow some of the techniques outlined in Chapter 5 on delivering your talk. In particular, the trio of presenting: Body language, voice and content.

I won't repeat those tips here, except to say (again) that you have to behave appropriately. The body language and the voice become very important when you are in a small group, particularly of people who know you. An auditorium of strangers won't know whether the man or woman with the stiff gestures and robotic voice behind the podium always sounds like that, but a handful of your associates in a room will. Your body language and voice need to appear natural and convey enthusiasm. I can offer you two tips on doing this:

Use Conversational Language

There is a difference between words which are written to be read, and those which are written to be spoken. This is one of the points I pick on when I help clients to develop their messages on a flipchart. When I write them, I write as I speak. (In fact I hope this whole book has a conversational style about it.) When they write them, they write them to be approved by lawyers. The lawyers then make the language more legalistic, and the end product sounds like a legal document, not a conversation.

Practise Saying the Words Out Loud

I worked in TV news for many years. If you ever have the chance to visit a TV newsroom, take it. In the scriptwriters' area you will find the writers, sitting behind computer screens muttering to themselves. They're not mad, they're speaking the scripts out loud. That's because that is how the words are to be delivered – verbally. You need to do the same. Key words, message tracks, story flows and Q&A documents on paper are all fine, but they are rarely conversational. Only by saying it out loud can you identify just the right words. If it's important, record yourself doing it. You can do this on your mobile phone or digital camera. Play it back and see how you can improve it.

Pulling It All Together

There are many tips and techniques in this book, and I would like to end it with a summary of the key ones.

- Plan your presentation with the audience in mind. Put yourself in their shoes. Think about what they know, want to know and need to know. Think about what you want to tell them. Consider where they are on the scale of 'emotional to intellectual'.
- Use audience-friendly language. Stress the benefits to them, not just the features. Don't confuse information with communication. Use appropriate language and body language.
- Start with the end in mind. Ask yourself what you want the audience to do, think, say or feel when they have heard you.
- Make things as simple as possible, but no simpler. Don't confuse information and communication.
- Use the grid system to prepare your talks. Anticipate objections and use attitude softeners to defuse them.
- The need to balance the risks and benefits, and convey them accurately, underpins every example of science communication. As the presenter you have a responsibility to your audience to *interpret* the data as well as *report* it.
- If you are giving a talk with slides, remember that PowerPoint is supporting you in your talk, not vice versa. Declutter your slides. Don't fill them with text. Don't read the slides. If you do, you become a soundtrack to the PowerPoint, and lose authority.

Ask yourself, 'Why am I presenting this data, rather than just sending the slides or details?'

- The agenda of journalists and scientists are often in conflict: Much journalism is presented as black and white, while much science is less definite. When planning a media interview, never underestimate their intelligence, but always underestimate their understanding.
- Develop memorable soundbites and summary points before the interview, then practise the interview with a colleague or media trainer.
- 'Off the record' is fraught with danger, and should only be used by experienced media spokespeople.
- During the interview, listen carefully to the whole question. When you answer, start with the conclusion then give the supporting information, not the other way round. Use the bridging technique where possible, but avoid sounding evasive.

A Personal Post Script

I said at the beginning of the book that science communication brings its own challenges. These are made more difficult by the increasing amount of what people far better qualified than I refer to as 'junk science', particularly on the internet. Overcoming the challenges is hard work, but I believe it is crucial. I often introduce myself at conferences by saying that my passion is communicating clearly about science, health and medicine. During this book I have tried to convey that to you, and I hope I have encouraged you to share my passion.

If you have stayed with it this far I am delighted, as I assume you have found it useful and, I hope interesting. One encouraging sign in recent years has been the increase in the number of real scientists who have started to take communication with non-scientists seriously. I have quoted a small number of them in the book. I hope in some small way the book has encouraged you to join them.

Index

Page numbers in **bold** refer to figures.

If you have found this book useful you may be interested in other titles from Gower

A Short Guide to Reputation Risk
Garry Honey
Paperback: 978-0-566-08995-4
Ebook: 978-0-566-08996-1

Advising Upwards
A Framework for Understanding and Engaging Senior Management Stakeholders
Edited by Lynda Bourne
Hardback: 978-0-566-09249-7
Ebook: 978-1-4094-3430-6

Healthcare Relationship Marketing
Strategy, Design and Measurement
Ira J. Haimowitz
Hardback: 978-0-566-09217-6
Ebook: 978-1-4094-3196-1

John Clare's Guide to Media Handling
John Clare
Paperback: 978-0-566-08698-4

Visit **www.gowerpublishing.com** and

- search the entire catalogue of Gower books in print
- order titles online at 10% discount
- take advantage of special offers
- sign up for our monthly e-mail update service
- download free sample chapters from all recent titles
- download or order our catalogue